Android UI Design with XML

Tutorial Book

by Camilus Raynaldo

Table of Contents

(Dedication Page)

I'd like to thank the Android developer community, Google, and the Open Handset Alliance for their vision, and also I want to thank all who have encouraged me in such a process to write a book about the Android user interface.

"UI was designed with help of ArtfulBits Android GUI Prototyping Stencil"
http://www.artfulbits.com/

About the Author

Camilus Raynaldo is a software developer with more than seven years of experience in .Net and java development.

Other fields of interest are: Web development (HTML/CSS, XML) and oracle database. He graduated from the "INUQUA" in 2004 with a Bachelor of Science degree in computer science
He writes android applications for personal use and works to develop an android community in his country.

He works as IT Engineering at ONI (Citizen Registration system in Haiti) since 2007.

Preface

Nowadays good User Interface is very essential for the success of any application in this competitive market.

There are a lot of Android books on the market, but most of them are aimed at professional users and non-zero, there are few books on the market that deals in depth about Android and sometimes puts the user in total confusion. The purpose of this book is to teach the user how to create user interfaces with XML which is much easier than Java and can achieve similar results.

Who should read this book?

This book is for anyone who wants to begin programming in android and learn how to create android user interfaces, without any knowledge of XML and Java

You can download example code used in this book at the author website (www.keroob.com)

Part I

Introducing Android

Chapter 1: Introducing Android

What is Android?

Android is a software stack for mobile devices that includes an operating system, middleware and key applications. The Android SDK provides the tools and APIs necessary to begin developing applications on the Android platform using the Java programming language.

Android is officially guided by the Open Handset Alliance but in reality Google leads the project and it's first truly open and comprehensive platform for mobile devices

Android relies on Linux version 2.6 for core system services such as security, memory management, process management, network stack, and driver model. The kernel also acts as an abstraction layer between the hardware and the rest of the software stack.

Android Sdk Features

- **Application framework** enabling reuse and replacement of components
- **Dalvik virtual machine** optimized for mobile devices
- **Integrated browser** based on the open source WebKit engine
- **Optimized graphics** powered by a custom 2D graphics library; 3D graphics based on the OpenGL ES 1.0 specification (hardware acceleration optional)
- **SQLite** for structured data storage
- **Media support** for common audio, video, and still image formats (MPEG4, H.264, MP3, AAC, AMR, JPG, PNG, GIF)
- **GSM Telephony** (hardware dependent)
- **Bluetooth, EDGE, 3G, and Wi-Fi** (hardware dependent)
- **Camera, GPS, compass, and accelerometer** (hardware dependent)

  Rich development environment including a device emulator, tools for debugging, memory and performance profiling, and a plugin for the Eclipse IDE

Android Application Architecture

The following diagram shows the major components of the Android operating system. Each section is described in more detail below.

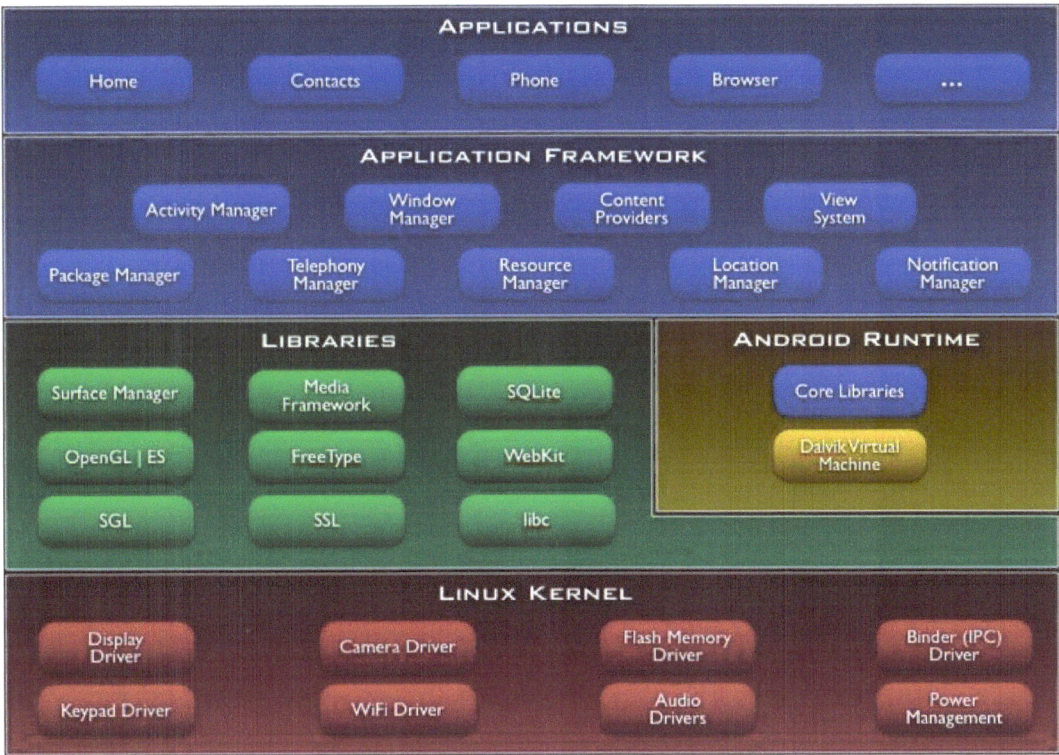

Application components

Application components are the essential building blocks of an Android application. Each component is a different point through which the system can enter your application. Not all components are actual entry points for the user and some depend on each other, but each one exists as its own entity and plays a specific role each one is a unique building block that helps define your application's overall behavior.

There are four different types of application components. Each type serves a distinct purpose and has a distinct lifecycle that defines how the component is created and destroyed.

Here are the four types of application components:

Activities

An *activity* represents a single screen with a user interface. For example, an email application might have one activity that shows a list of new emails, another activity to compose an email, and another activity for reading emails. Although the activities work together to form a cohesive user experience in the email application, each one is independent of the others. As such, a different application can start any one of these activities (if the email application allows it). For example, a camera application can start the activity in the email application that composes new mail, in order for the user to share a picture.

Services

A *service* is a component that runs in the background to perform long-running operations or to perform work for remote processes. A service does not provide a user interface. For example, a service might play music in the background while the user is in a different application, or it might fetch data over the network without blocking user interaction with an activity. Another component, such as an activity, can start the service and let it run or bind to it in order to interact with it.

Content providers

A *content provider* manages a shared set of application data. You can store the data in the file system, a SQLite database, on the web, or any other persistent storage location your application can access. Through the content provider, other applications can query or even modify the data (if the content provider allows it). For example, the Android system provides a content provider that manages the user's contact information. As such, any application with the proper permissions can query part of the content provider (such as ContactsContract.Data) to read and write information about a particular person.

Content providers are also useful for reading and writing data that is private to your application and not shared

Broadcast receivers

A broadcast receiver is a component that responds to system-wide broadcast announcements. Many broadcasts originate from the system—for example, a broadcast announcing that the screen has turned off, the battery is low, or a picture was captured. Applications can also initiate broadcasts, for example, to let other applications know that some data has been downloaded to the device and is available for them to use. Although broadcast receivers don't display a user interface, they may create a status bar notification to alert the user when a broadcast event occurs. More commonly, though, a broadcast receiver is just a "gateway" to other components and is intended to do a very minimal amount of work. For instance, it might initiate a service to perform some work based on the event.

A unique aspect of the Android system design is that any application can start another application's component. For example, if you want the user to capture a photo with the device camera, there's probably another application that does that and your application can use it, instead of developing an activity to capture a photo yourself. You don't need to incorporate or even link to the code from the camera application. Instead, you can simply start the activity in the camera application that captures a photo. When complete, the photo is even returned to your application so you can use it. To the user, it seems as if the camera is actually a part of your application.

When the system starts a component, it starts the process for that application (if it's not already running) and instantiates the classes needed for the component. For example, if your application starts the activity in the camera application that captures a photo, that activity runs in the process that belongs to the camera application, not in your application's process. Therefore, unlike applications on most other systems, Android applications don't have a single entry point (there's no *main()* function, for example).

Because the system runs each application in a separate process with file permissions that restrict access to other applications, your application cannot directly activate a component from another application. The Android system, however, can. So, to activate a component in another application, you must deliver a message to the system that specifies your *intent* to start a particular component. The system then activates the component for you.

Installing Android

The development environment used for the sample applications in this book includes:

- Microsoft Windows XP/Vista/7
- Java Development Kit (JDK) version 6
- The Android SDK available for download at the google developer website:
  http://developer.android.com/sdk/index.html
- Eclipse IDE (Integrated Development Environment) version 3.7
- ADT plugin for Eclipse

Then create a directory *"My Folder"*, extract the content of the Eclipse file (Zip) and copy the JDK and the Android SDK installer.

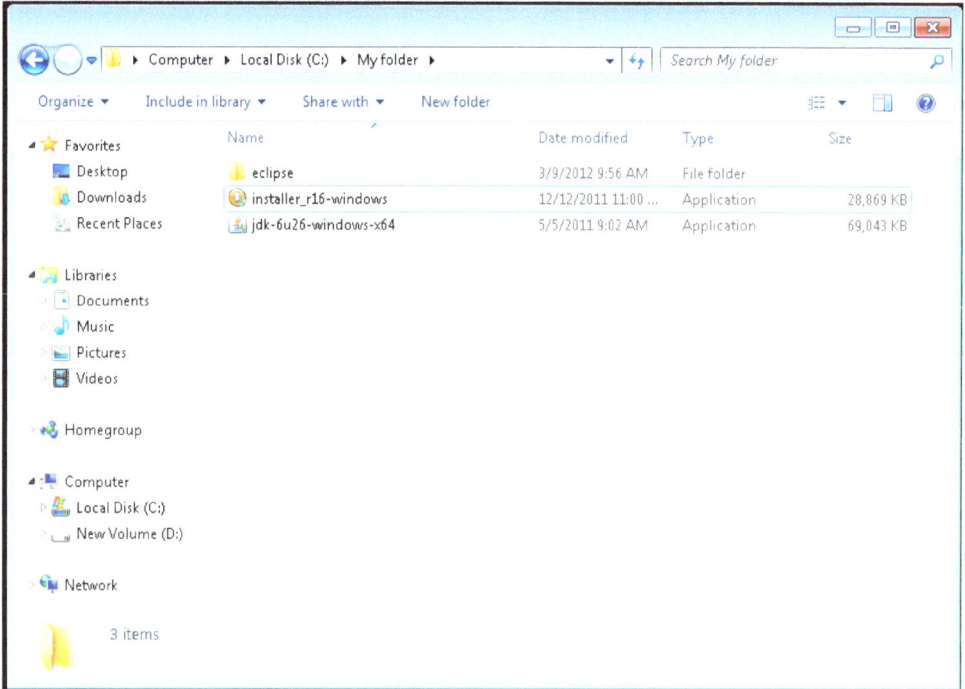

Supported Operating Systems

Windows XP (32-bit), Vista (32- or 64-bit), or Windows 7 (32- or 64-bit)

Mac OS X 10.5.8 or later (x86 only)

Linux (tested on Ubuntu Linux, Lucid Lynx) ∘GNU C Library (glibc) 2.7 or later is required.

On Ubuntu Linux, version 8.04 or later is required.

64-bit distributions must be capable of running 32-bit applications.

Installing the JDK

1- Double click the *jdk-6u26-windows-x64* to launch the Installation Wizard

2- Click **Next**

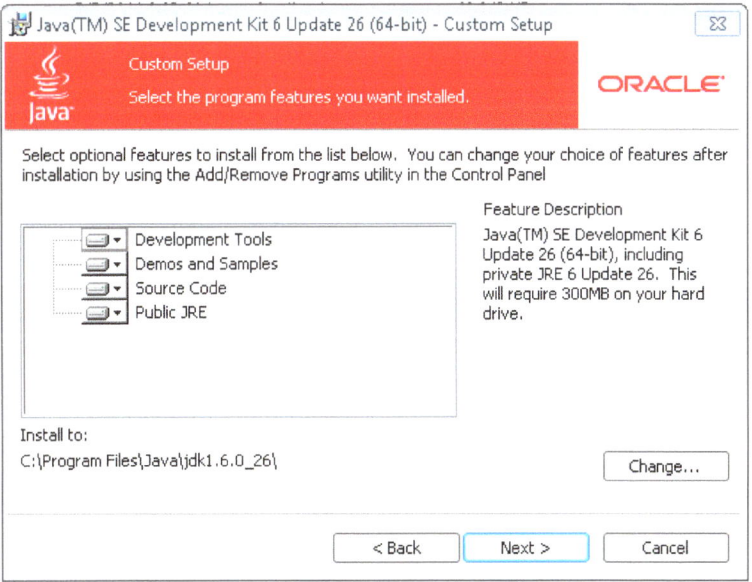

3- Click **Next** to choose your installation folder

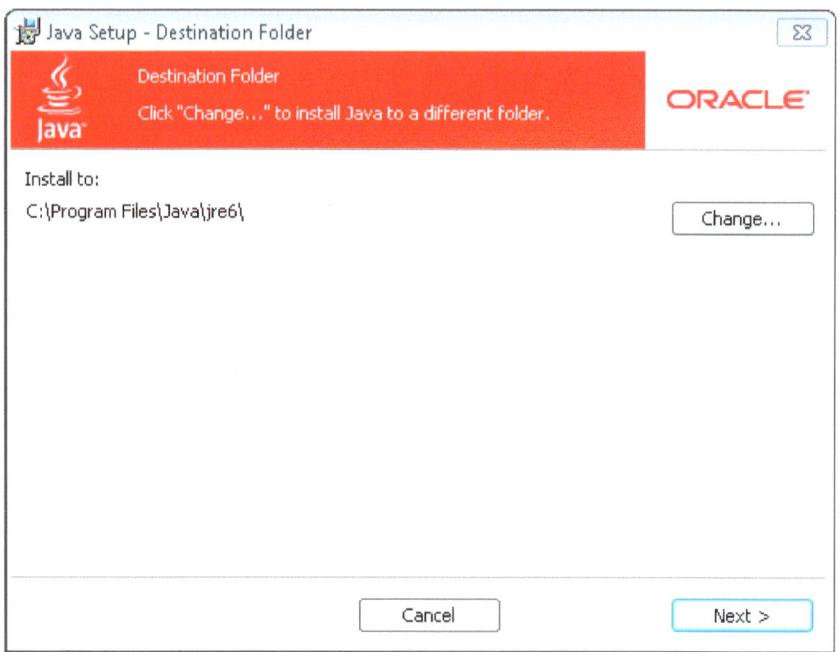

4- Click **Next** to begin the installation

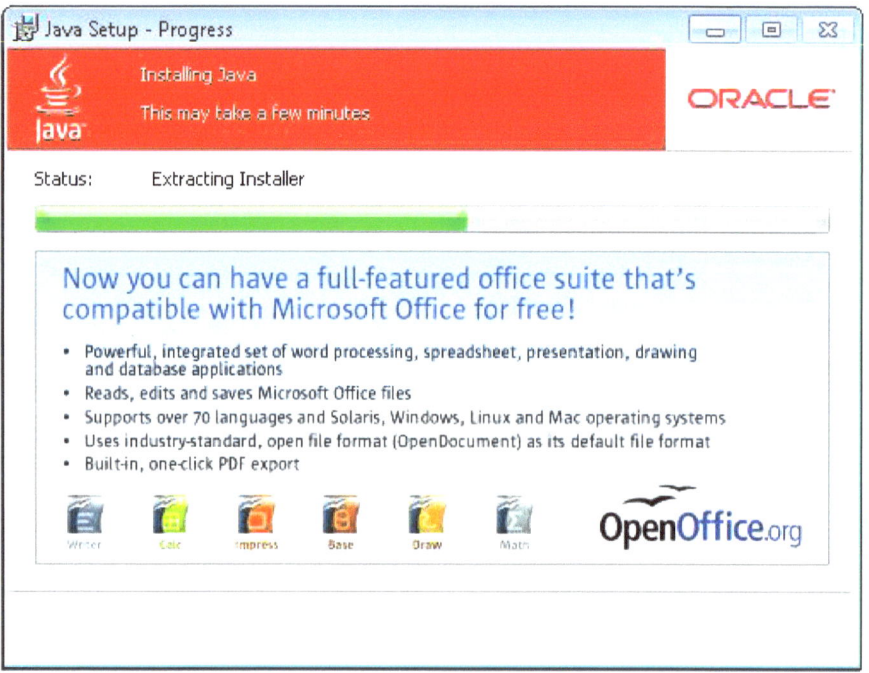

5- Once the progression ended, the Product Registration window appears

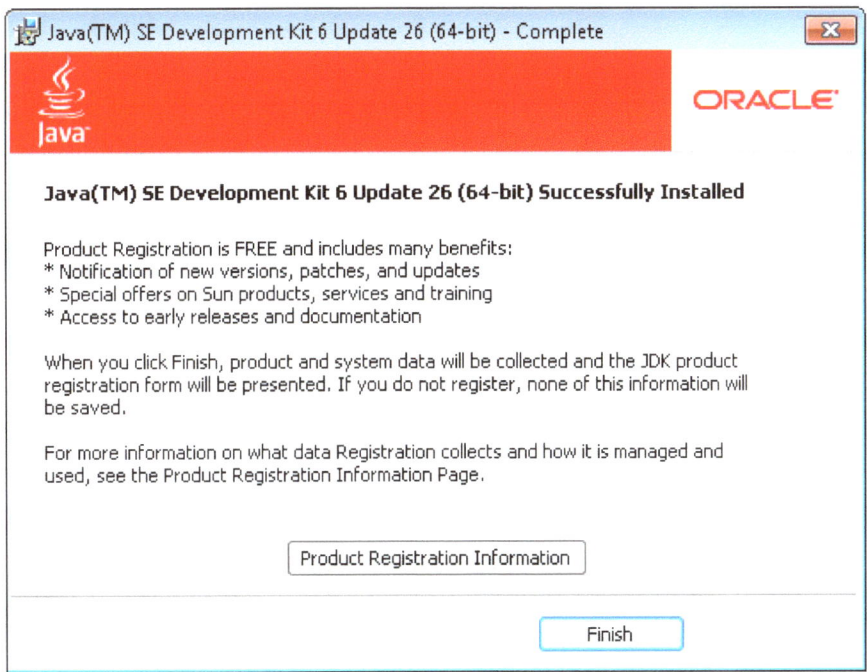

6- Click **Finish**

Installing the SDK

1- Double click *installer_r16-windows,* the Welcome to the Android SDK Tools window appears

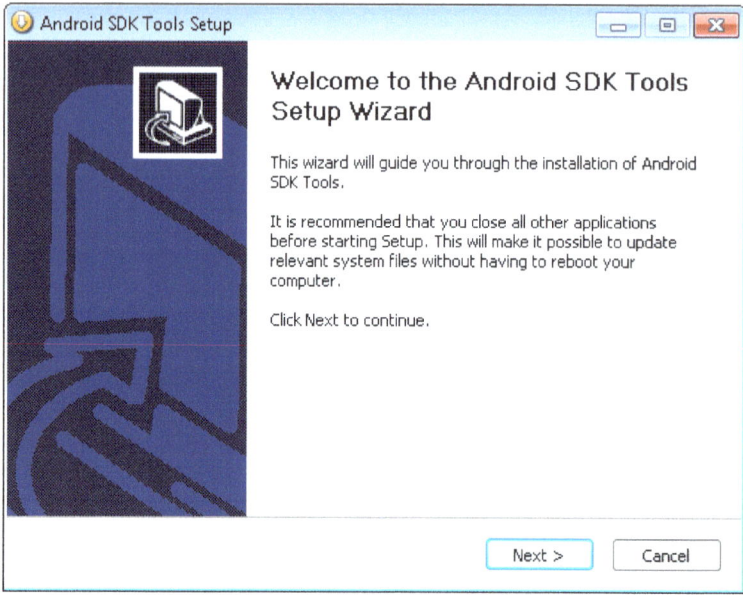

2- Click **Next**

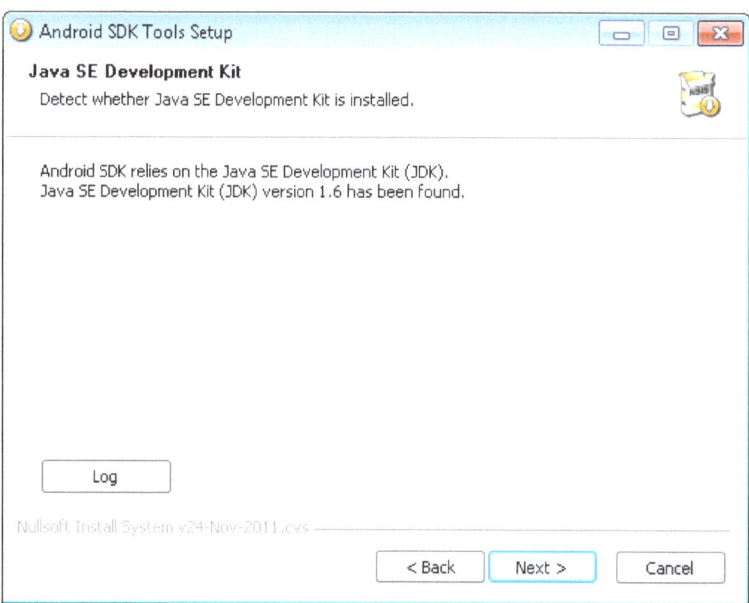

3- Click **Next** and choose the Install Location

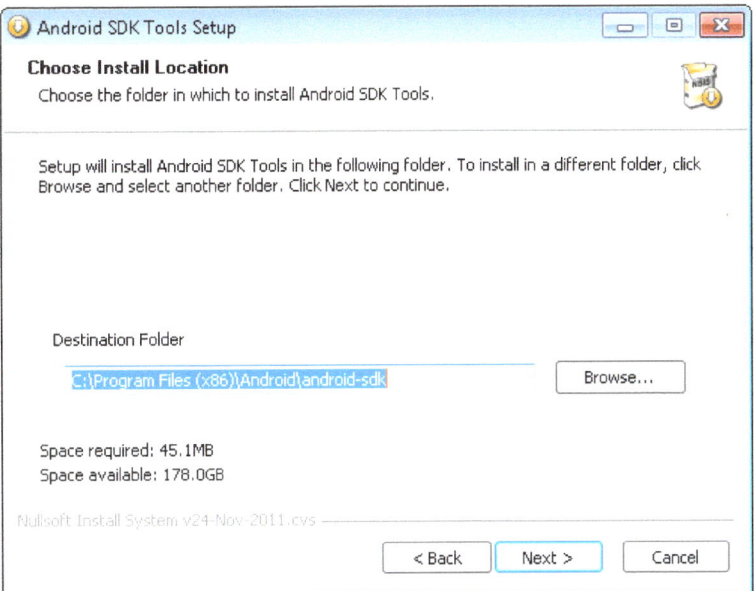

4- Click **Next**

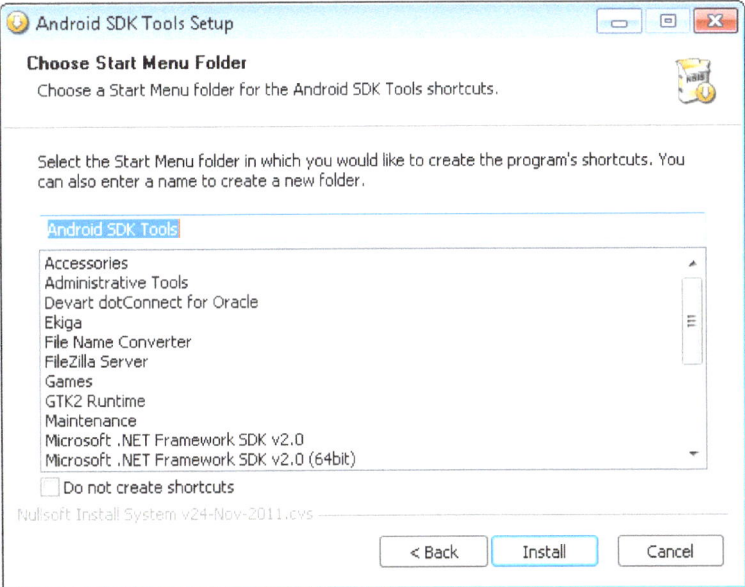

5- Click **Install**

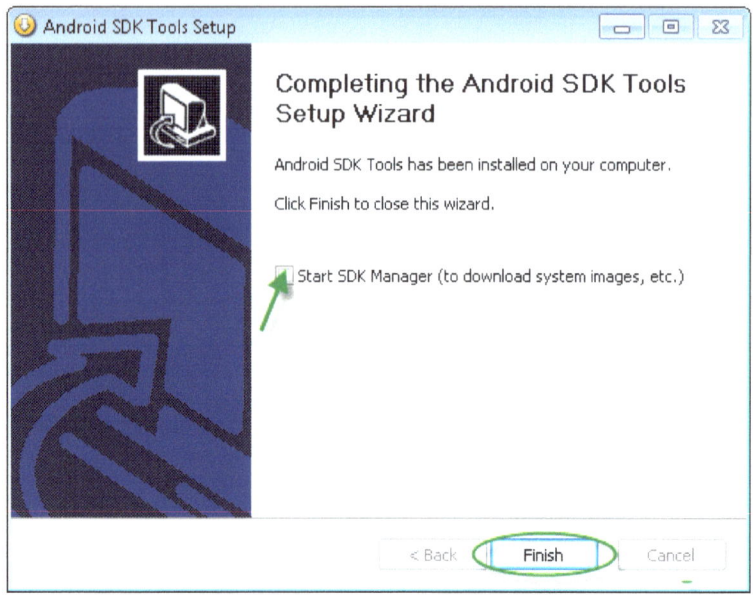

6- Then uncheck **"Start SDK Manager"** cause you will install it later and click **Finish**

Note: You can choose not to install the SDK Manager, you can do it later in the book during the installation of platforms and other components

Installing Eclipse

To install Eclipse you must first download and extract the file in *"My Folder"*, then create a shortcut of the Ecplise executable on the desktop

Installing the ADT Plugin for Eclipse

1- Double click on the **Eclipse** icon to start Eclipse for the first time

The Workspace Launcher window appears, and allows you to specify a workspace directory; You can use the default Workspace.

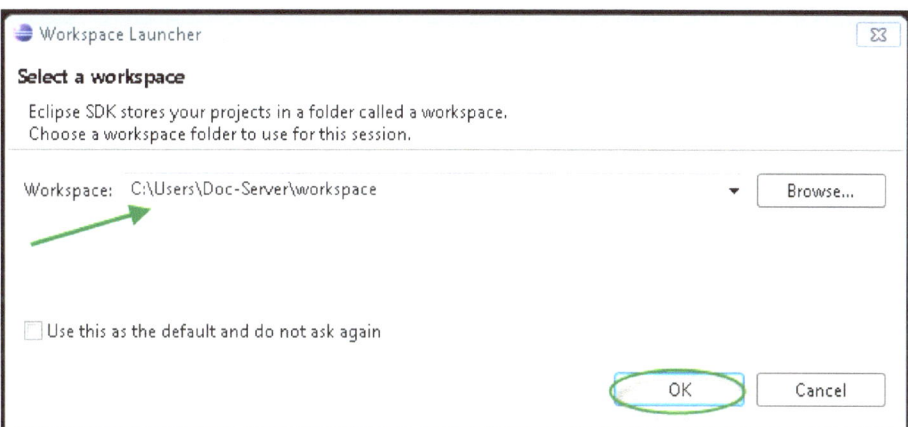

2- Click **OK**, then the Eclipse welcome screen appears

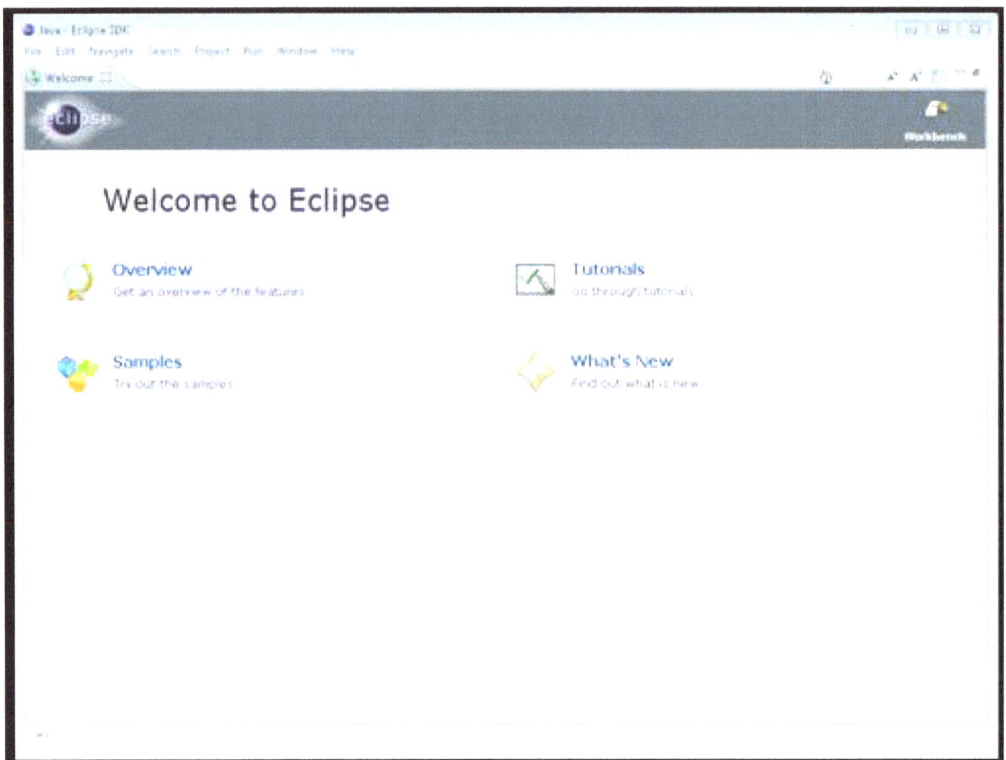

3- select **Help > Install New Software**

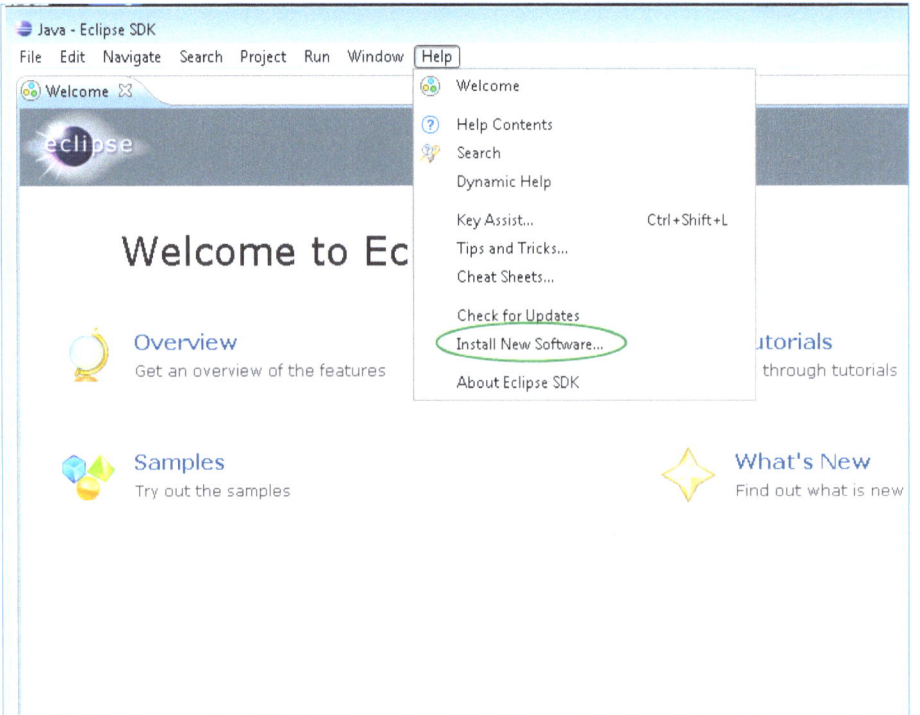

4- Click **Add**, in the top-right corner.

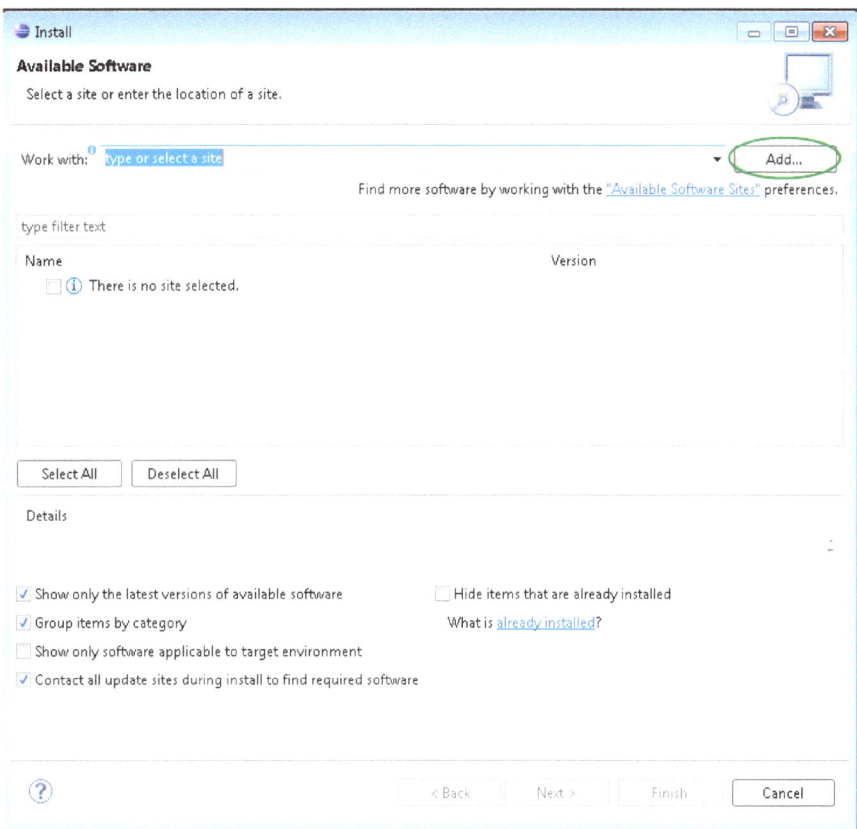

5- In the Add Repository dialog that appears, enter **"ADT Plugin"** for the *Name* and "https://dl-ssl.google.com/android/eclipse/ " for the *Location*:

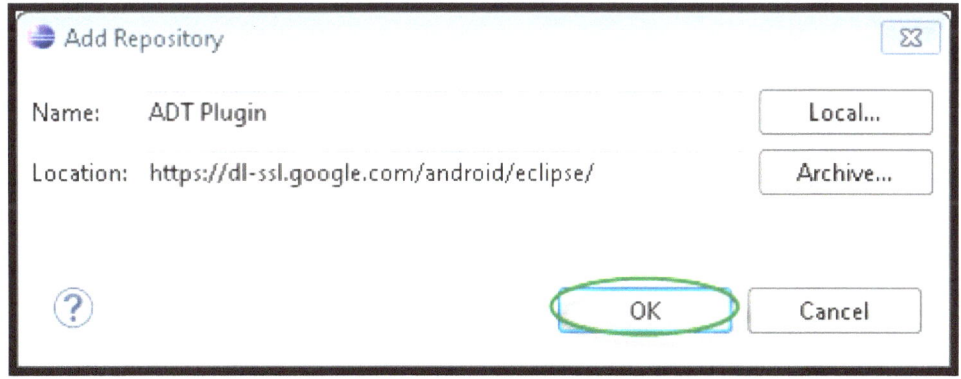

6- Click **OK**

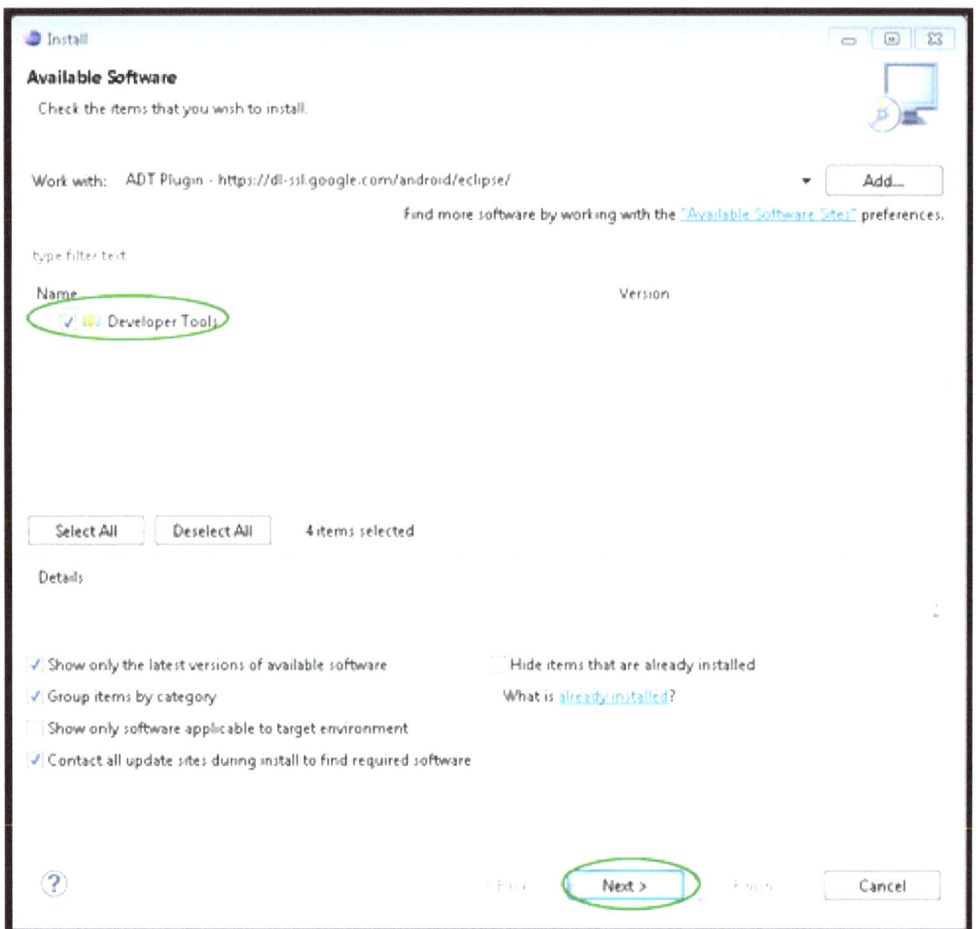

7- In the Available Software dialog, select the checkbox next to Developer Tools and click **Next**

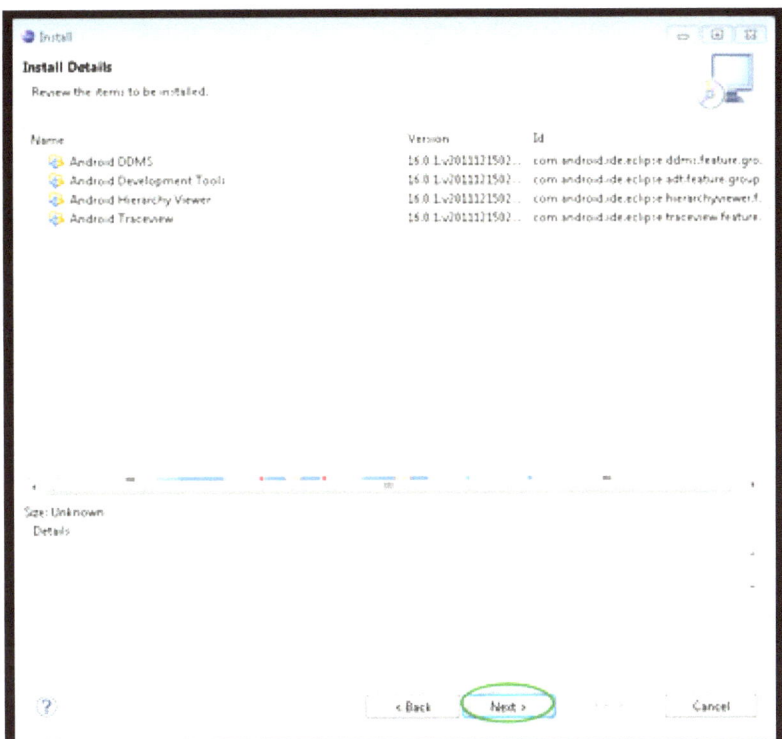

8- In the next window, you will see a list of items to be downloaded. Click **Next**.

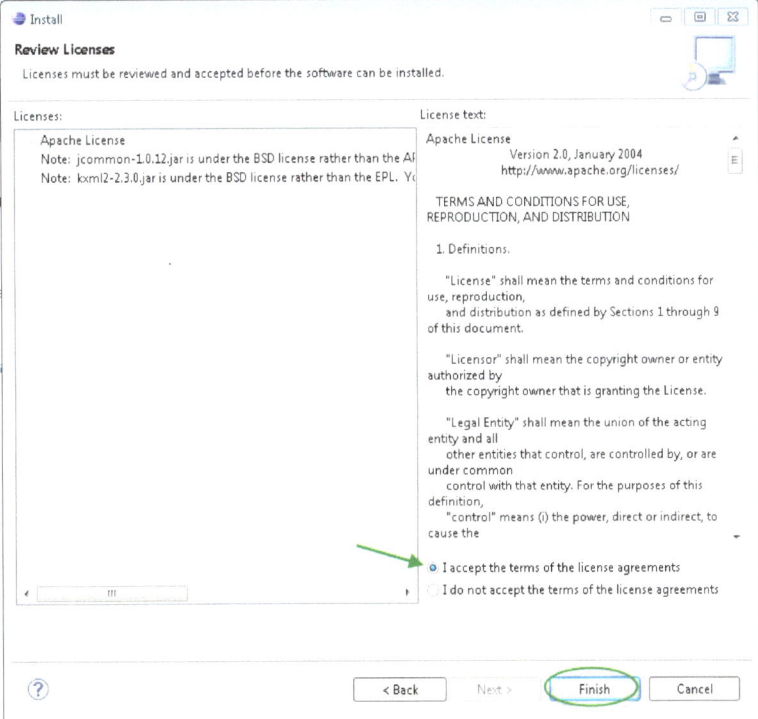

9- Read and accept the license agreements, then click **Finish**.

> **Note:** If you get a security warning saying that the authenticity or validity of the software can't be established, click **OK**.

10-A new window appears which shows the progress of the installation

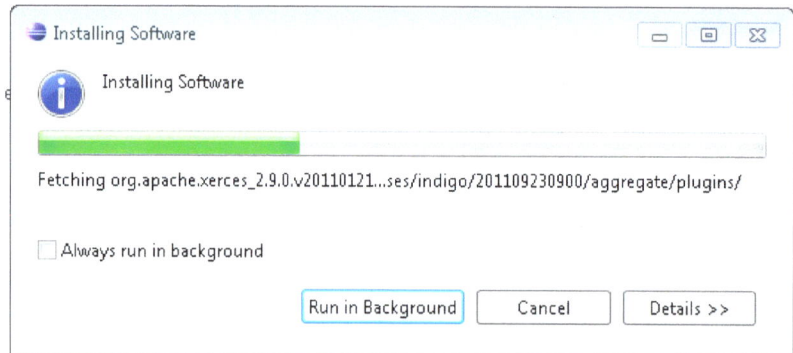

11-When the installation completes, Restart Eclipse.

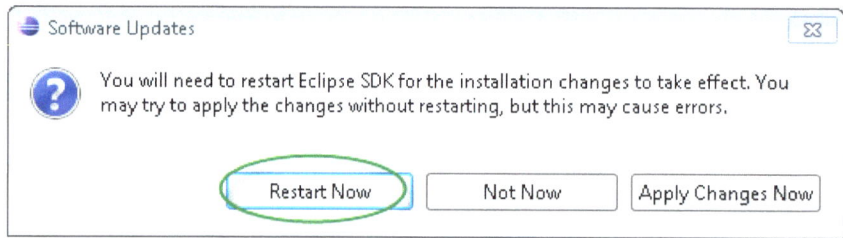

Adding platforms and other components

Now you have finished with the installation, add platforms and other components.

1- Select **HELP > Install New Software...**

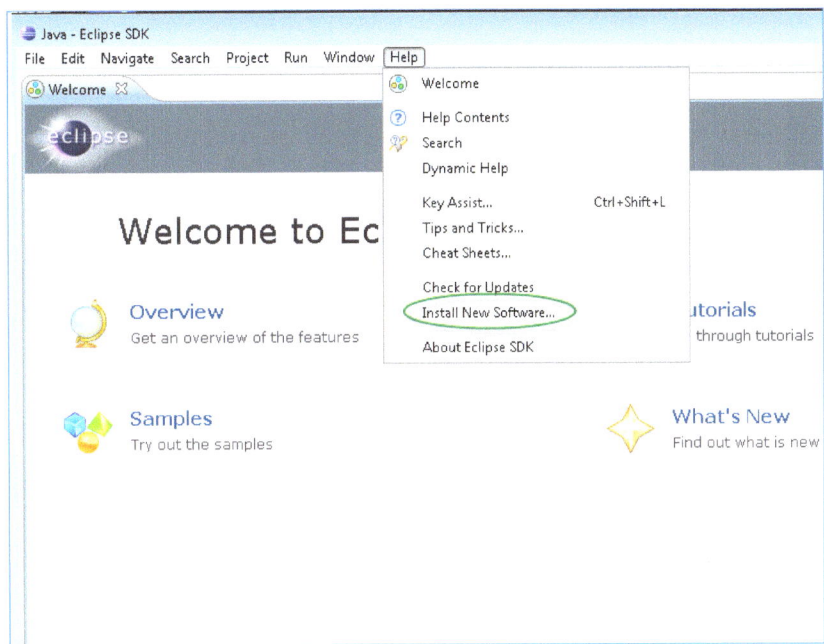

2- A new Window appears allowing you to select the platforms and components to install

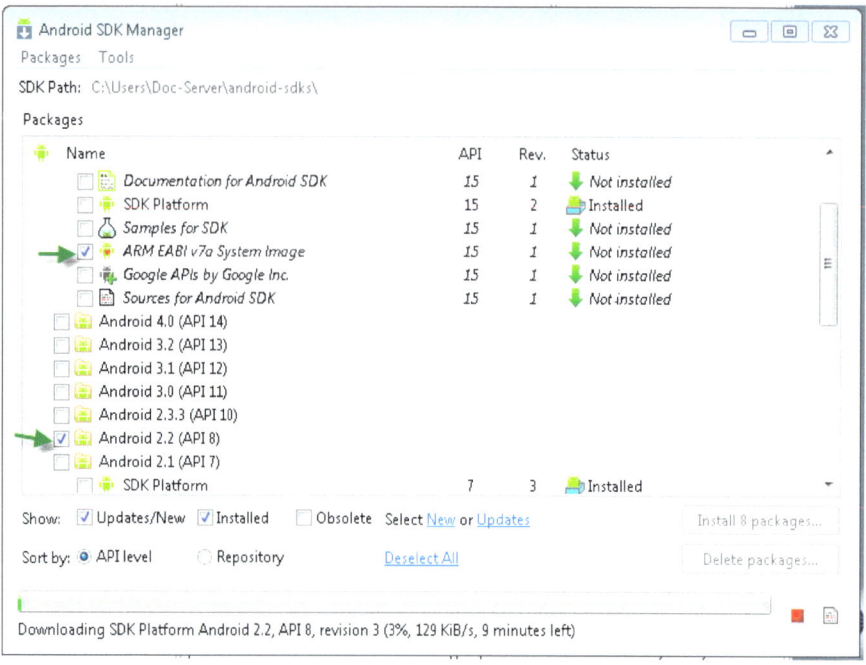

Configuring the ADT Plugin

Now you must modify the ADT preferences in Eclipse to point to the Android SDK directory:

1- Select **Window** > **Preferences...** to open the Preferences panel

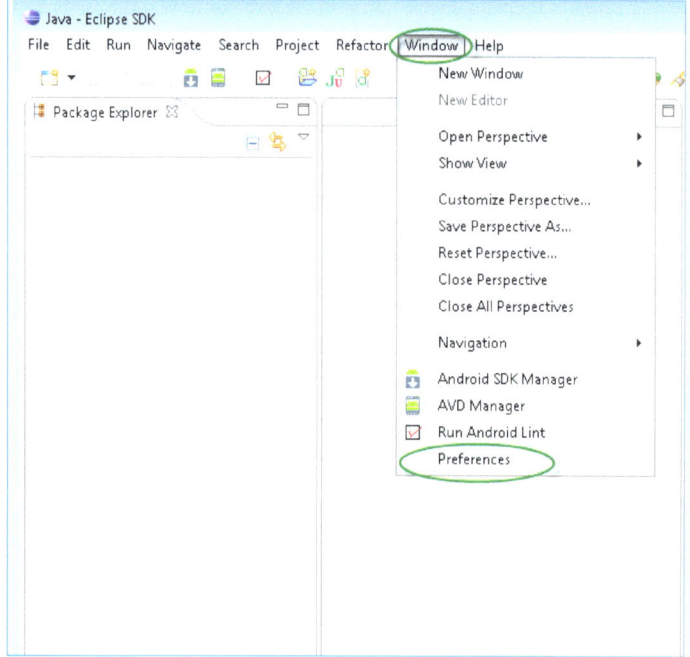

2- Select **Android** from the left panel.

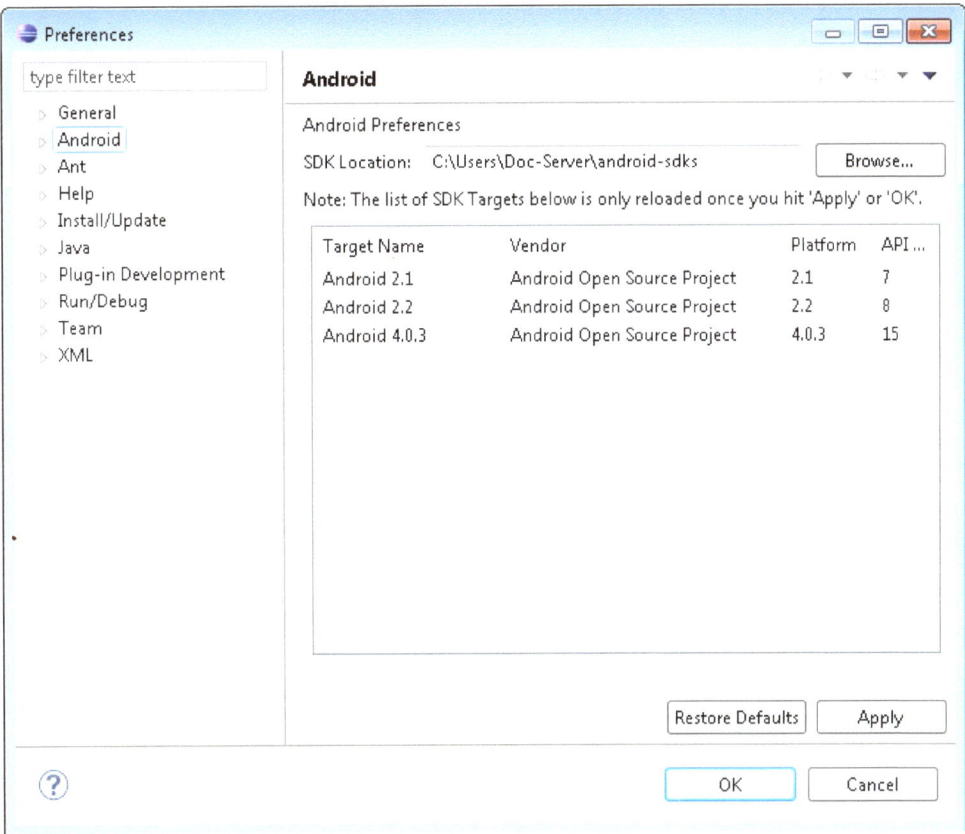

3- For the SDK Location in the main panel, click **Browse...** and locate your downloaded SDK directory.

4- Click **Apply**, then click **OK**

Create our first project

1- Double click on the eclipse icon

2- Choose **File > New > Other** to create a new project

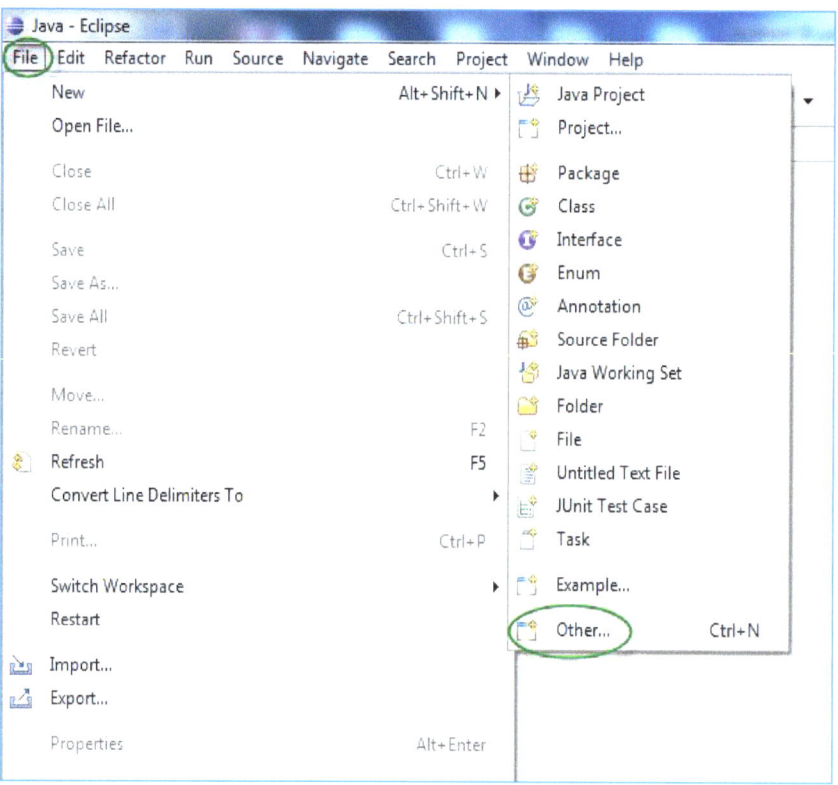

3- Select **Android Project** and click Next

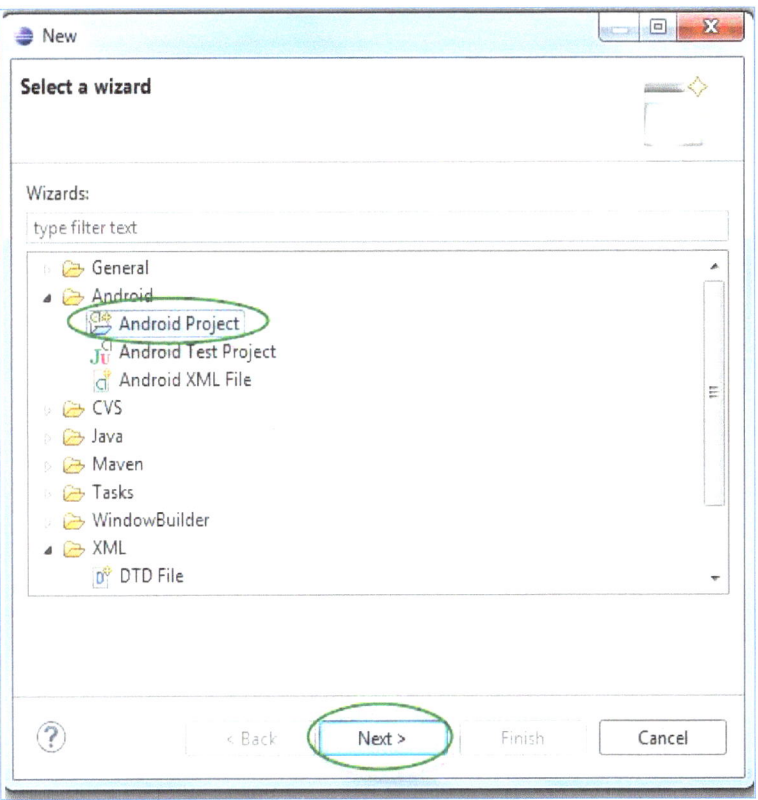

4- In Project Name enter **Hello Android**

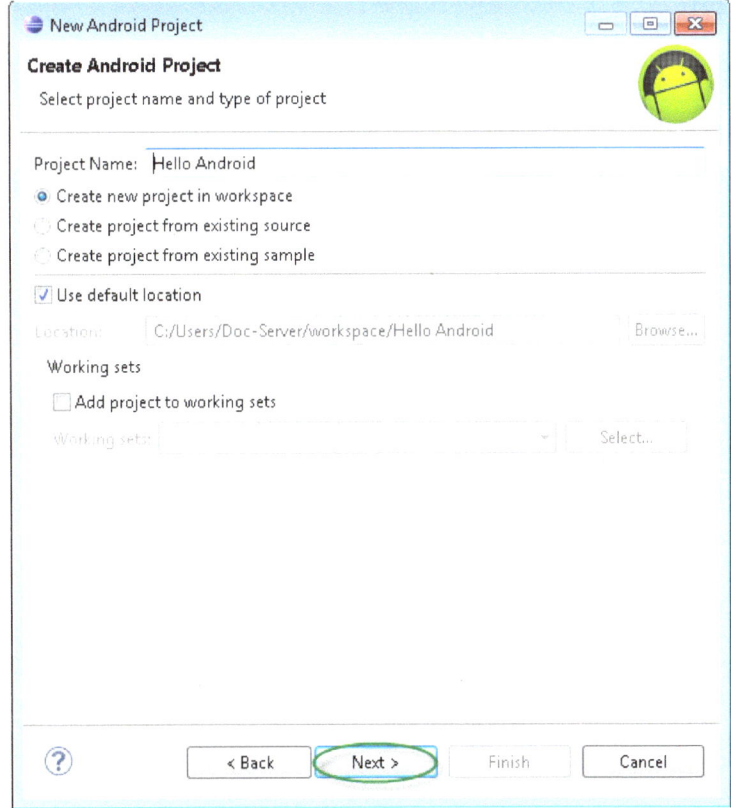

5- Select the **Build Target** (Android 2.2)

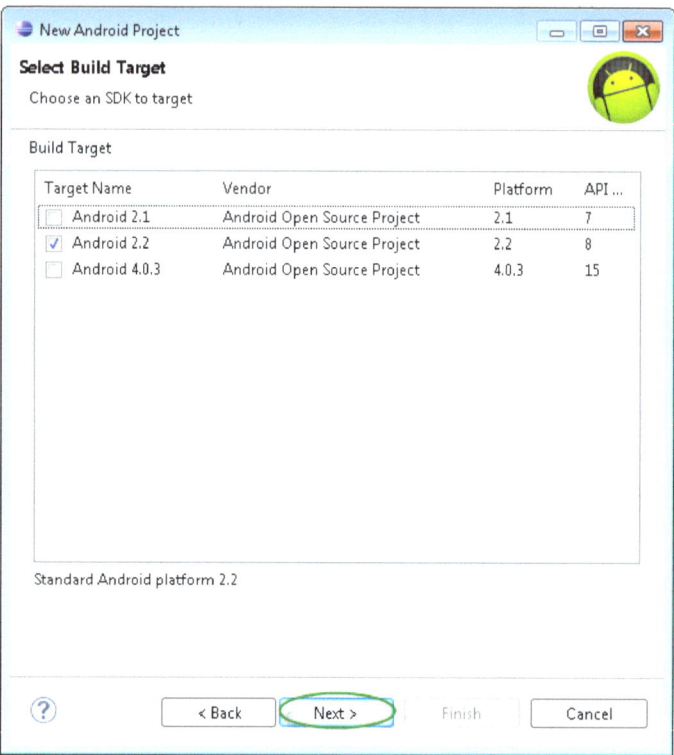

6- In **Package Name** enter "com.UInterface.book.helloandroid" and "8" in **Minimum SDK**

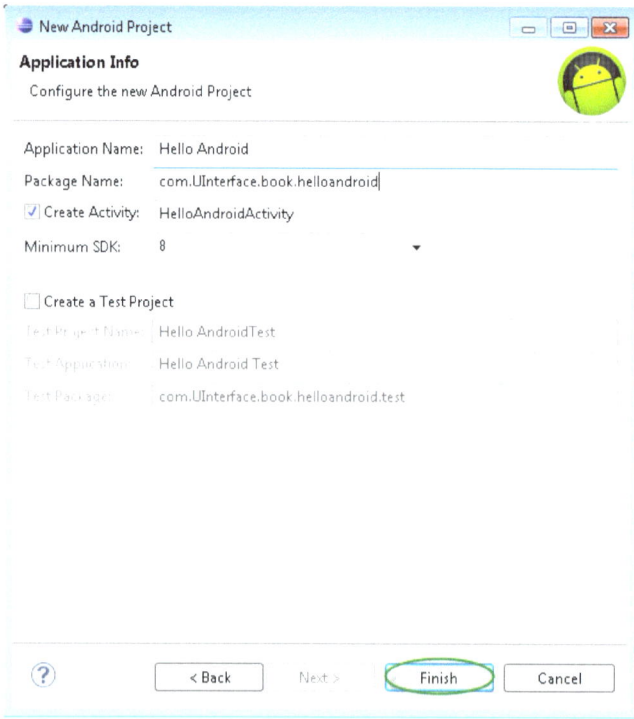

7- Click **Finish** button

Exploring your first android project files in the Package Explorer

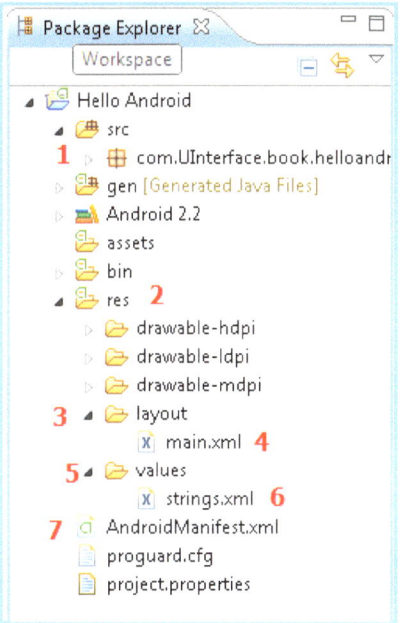

HelloAndroidActivity.java   1

Double click in HelloAndroidActivity.java and you will only see java code

```
package com.UInterface.book.helloandroid;

import android.app.Activity;
import android.os.Bundle;

public class HelloAndroidActivity extends Activity {
    /** Called when the activity is first created. */
    @Override
    public void onCreate(Bundle savedInstanceState) {
        super.onCreate(savedInstanceState);
        setContentView(R.layout.main);
    }
}
```

Resources folder:

Contains all the resources used by your Android application such as drawable files, layout files, and string values   2

Layout folder Res:

Contains an XML file used to represent the user interface of your Android application **3**

Main.xml **4**

This file allow you to create the GUI of your application

```xml
<?xml version="1.0" encoding="utf-8"?>
<LinearLayout xmlns:android="http://schemas.android.com/apk/res/android"
    android:layout_width="fill_parent"
    android:layout_height="fill_parent"
    android:orientation="vertical" >

    <TextView
        android:layout_width="fill_parent"
        android:layout_height="wrap_content"
        android:text="@string/hello" />

</LinearLayout>
```

Values folder:

For XML files that are compiled into many kinds of resource, resources written to XML files in this folder are not referenced by the file name. **5**

Strings Res:

Used to display text in your application **6**

```xml
<?xml version="1.0" encoding="utf-8"?>
<resources>
  <string name="hello">Hello World, HelloAndroidActivity!</string>
  <string name="app_name">Hello Android</string>
</resources>
```

AndroidManifest.xml **7**

The control file that describes the nature of the application and each of its components, for instance, it describes certain qualities about the activities, services, intent receivers, and content providers; what permissions are requested; what external libraries are needed; what device features are required, what API Levels are supported or required; and others.

```xml
<?xml version="1.0" encoding="utf-8"?>
<manifest xmlns:android="http://schemas.android.com/apk/res/android"
    package="com.androui.hello"
    android:versionCode="1"
    android:versionName="1.0">
    <uses-sdk android:minSdkVersion="8" />

    <application android:icon="@drawable/icon"
android:label="@string/app_name">
```

```
            <activity android:name=".HelloAndroidActivity"
                    android:label="@string/app_name">
                <intent-filter>
                    <action android:name="android.intent.action.MAIN" />
                    <category android:name="android.intent.category.LAUNCHER" />
                </intent-filter>
            </activity>

        </application>
</manifest>
```

Gradually as you move forward in this book you will study in detail these various files.

Now compile your first application, above all create your first emulator.

Creating an android virtual device

The Menu Bar of the Eclipse software looks like the following:

You can create as many AVDs as you would like to test on. It is recommended that you test your applications on all API levels higher than the target API level for your application.

1- In the eclipse menu select **window > AVD Manager,** or click AVD Manager icon in the Eclipse toolbar

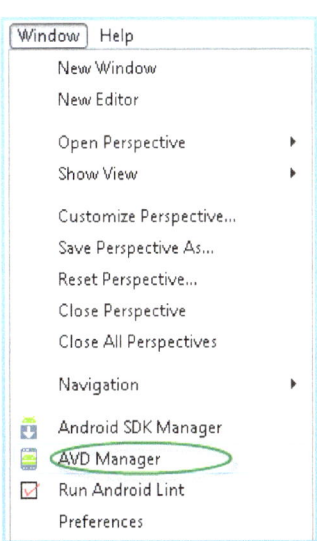

2- In the Android *Virtual Devices* Manager. Click **New** to create a new AVD. The **Create New AVD** dialog appears.

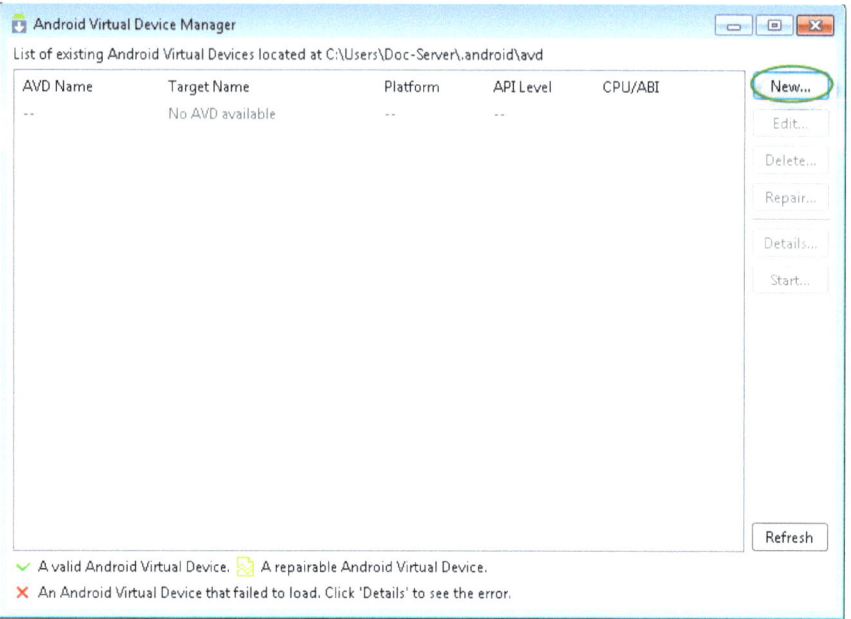

Fill in the details for the AVD.

Name = YourAVD, **platform target** = Android 2.2-API Level8, **SD card size**, and a **skin** (HVGA is default).

3- Click **Create AVD**.

Running an Android application

1- To run (or debug) your application, select **Run** > **Run** (or **Run** > **Debug**) from the Eclipse menu bar, or right click the project in the package explorer **Run As** > **Android Application**

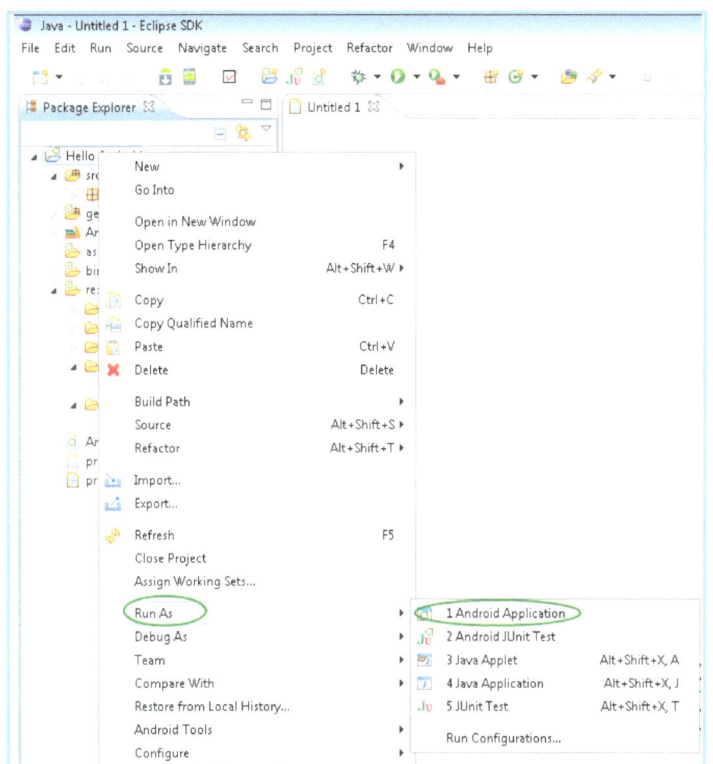

2- Click and drag the padlock to the right

3- Result should look like this:

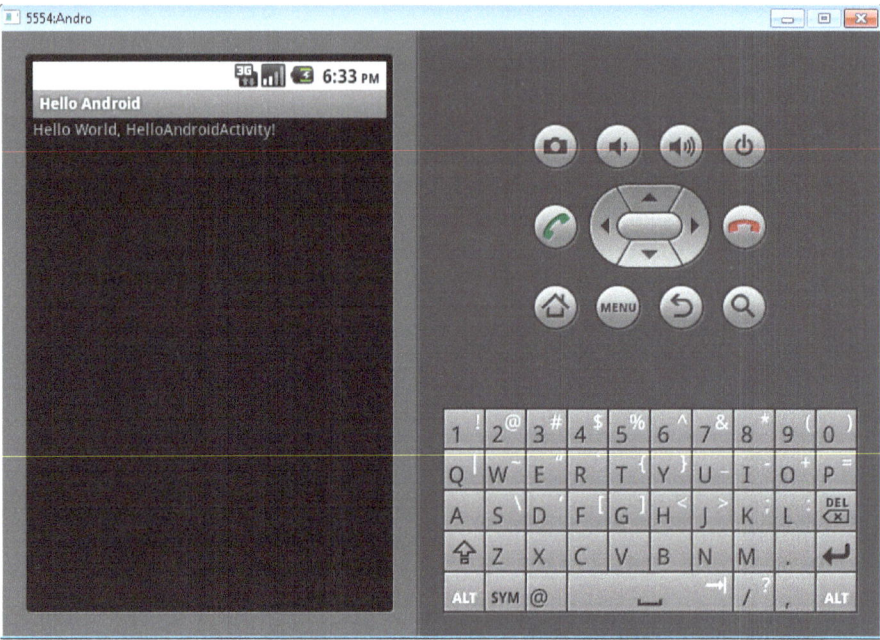

Part II

Android User Interface

Chapter 2: Android XML layout elements

XML Layouts

A layout is the architecture for the user interface in an Activity. It defines the layout structure and holds all the elements that appear to the user *(buttons, images, Textview etc)*. You can declare your layout in two ways:

1- Declare UI elements in XML
2- Instantiate layout elements at runtime

The advantage to declaring your UI in XML is that it enables you to better separate the presentation of your application from the code that controls its behavior. Your UI descriptions are external to your application code, which means that you can modify or adapt it without having to modify your source code and recompile. For example, you can create XML layouts for different screen orientations, different device screen sizes, and different languages. Additionally, declaring the layout in XML makes it easier to visualize the structure of your UI, so it's easier to debug problems.

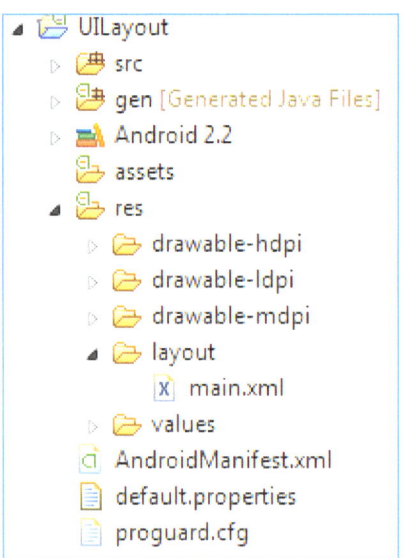

This book focuses on teaching you how to declare your layout in XML.

> Note: The **ADT Plugin for Eclipse** offers a layout preview of your XML with the XML file opened, select the **Layout** tab.

Write the XML

Using Android's XML vocabulary, you can quickly design UI layouts and the screen elements they contain, in the same way you create web pages in HTML with a series of nested elements.

Each layout file must contain exactly one root element, which must be a View or ViewGroup object. Once you've defined the root element, you can add additional layout objects or widgets as child elements to gradually build a View hierarchy that defines your layout.

Create an XML file in eclipse

1- Choose **New** > **Other**

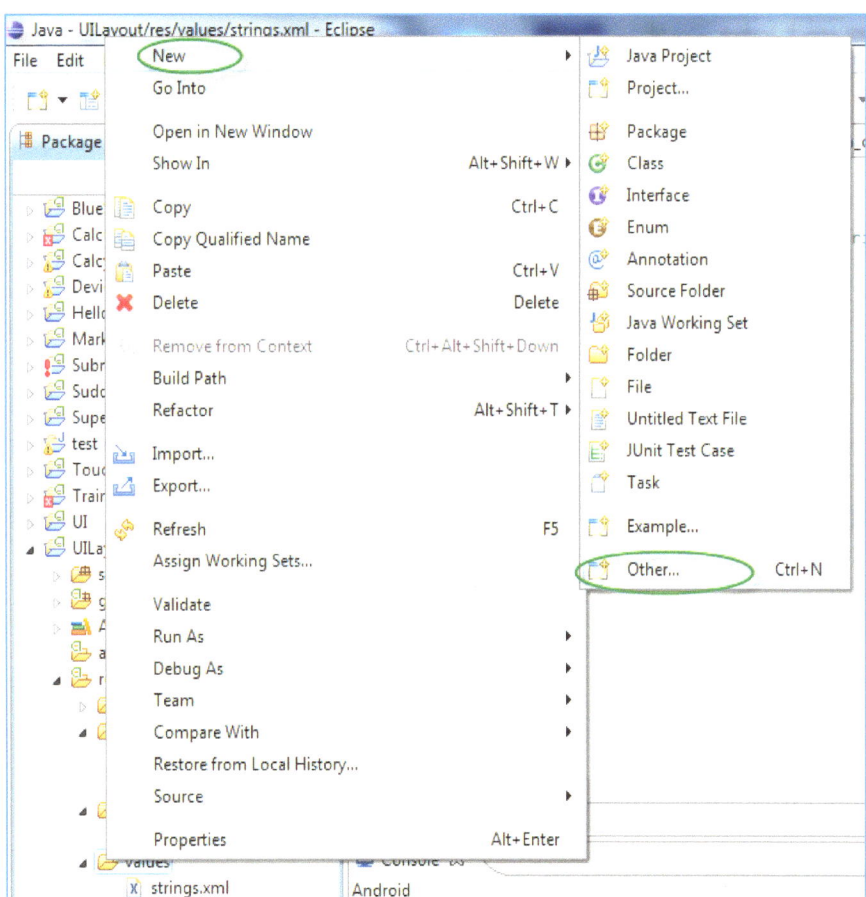

2- A new window appears, select **Android XML File**

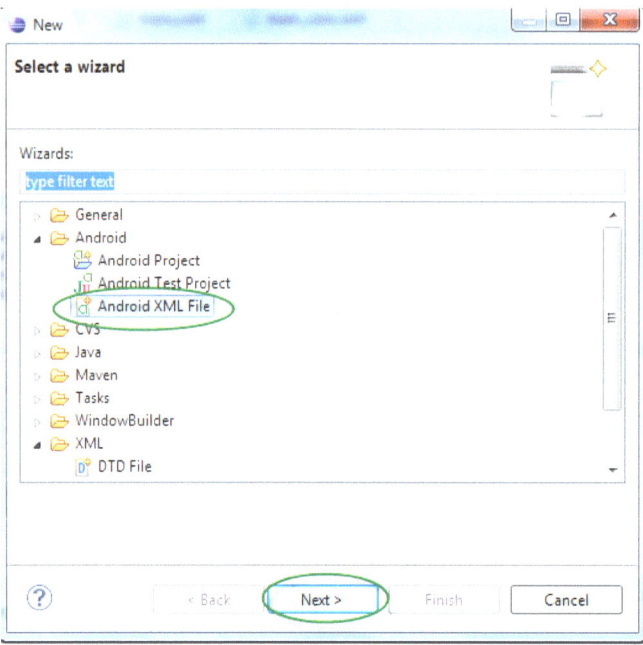

3- Click **Next**

4- In the « **New Android XML file** » window provide a filename **1** and select the type of resource to create **2**

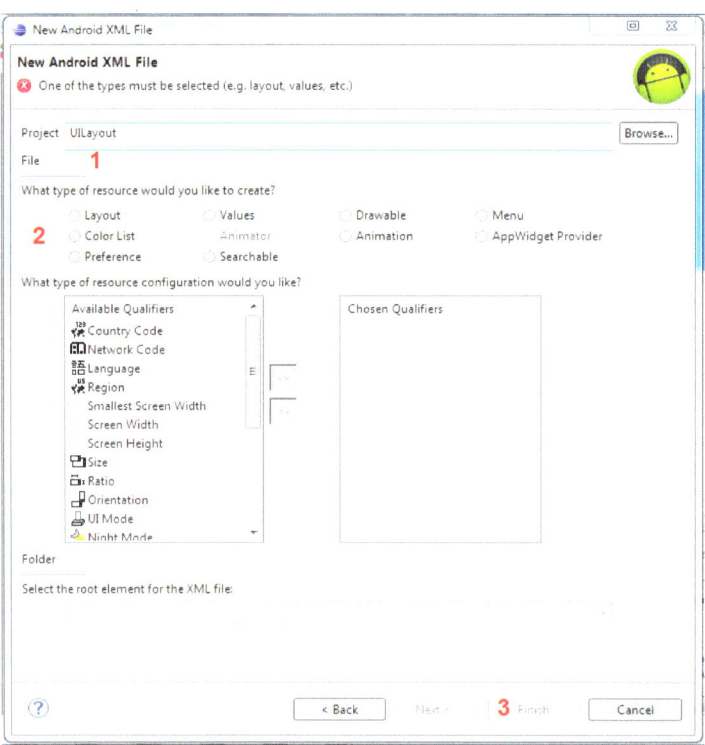

5- Click **Finish** button **3**

Attributes

ID

Any View object may have an integer ID associated with it, to uniquely identify the View within the tree. When the application is compiled, this ID is referenced as an integer, but the ID is typically assigned in the layout XML file as a string, in the id attribute. This is an XML attribute common to all View objects (defined by the **View** class) and you will use it very often. The syntax for an ID, inside an XML tag is:

android:id="@+id/bouton"

(@): indicates that the XML parser should parse and expand the rest of the ID string and identify it as an ID resource.

(+): means that this is a new resource name that must be created and added to our resources (in the R.java file).

There are a number of other ID resources that are offered by the Android framework. When referencing an Android resource ID, you do not need the plus-symbol, but must add the android package namespace, like so:

android:id="@android:id/empty"

An ID need not be unique throughout the entire tree, but it should be unique within the part of the tree you are searching (which may often be the entire tree, so it's best to be completely unique when possible).

Layout Parameters

XML layout attributes named **layout_something** define layout parameters for the View that is appropriate for the ViewGroup in which it resides.

Every ViewGroup class implements a nested class that extends ViewGroup.LayoutParams. This subclass contains property types that define the size and position for each child view, as appropriate for the view group. As you can see in figure 1, the parent view group defines layout parameters for each child view (including the child view group).

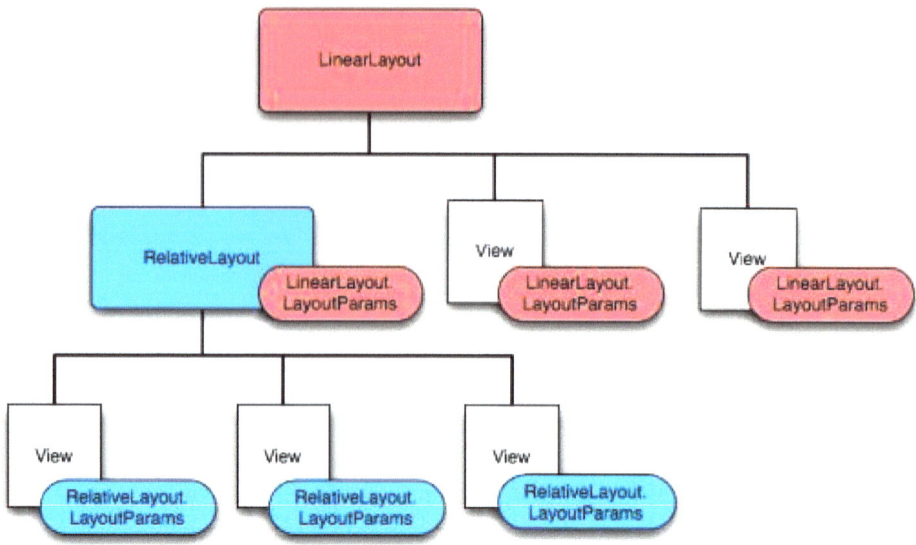

Note that every LayoutParams subclass has its own syntax for setting values. Each child element must define LayoutParams that are appropriate for its parent, though it may also define different LayoutParams for its own children.

All view groups include a width and height (**layout_width** and **layout_height**), and each view is required to define them. Many LayoutParams also include optional margins and borders.

XML Attributes

Attribute Name	Related Method	Description
android:layout_height		Specifies the basic height of the view.
android:layout_width		Specifies the basic width of the view.

wrap_content tells your view to size itself to the dimensions required by its content
fill_parent (renamed *match_parent* in API Level 8) tells your view to become as big as its parent view group will allow

Dp: Density-independent Pixels - an abstract unit that is based on the physical density of the screen. These units are relative to a 160 dpi (dots per inch) screen, so *160dp is always one inch* regardless of the screen density. The ratio of dp-to-pixel will change with the screen density, but not necessarily in direct proportion. You should use these units when specifying view dimensions in your layout, so the UI properly scales to render at the same actual size on different screens. (The compiler accepts both "dip" and "dp", though "dp" is more consistent with "sp".)

Sp: Scale-independent Pixels - this is like the dp unit, but it is also scaled by the user's font size preference. It is recommend you use this unit when specifying font sizes, so they will be adjusted for both the screen density and the user's preference.

Pt: Points - 1/72 of an inch based on the physical size of the screen.

Px: Pixels - corresponds to actual pixels on the screen. This unit of measure is not recommended because the actual representation can vary across devices; each device may have a different number of pixels per inch and may have more or fewer total pixels available on the screen.

Mm: Millimeters - based on the physical size of the screen.

In: Inches - based on the physical size of the screen.

In general, specifying a layout width and height using absolute units such as pixels is not recommended. Instead, using relative measurements such as density-independent pixel units *(dp), **wrap_content**, or **fill_parent**,* is a better approach, because it helps ensure that your application will display properly across a variety of device screen sizes.

Chapter 3: Android GUI widget

The widget package contains (mostly visual) UI elements to use on your Application screen. You can design your own

To create your own widget, extend View or a subclass. To use your widget in layout XML, there are two additional files for you to create. Here is a list of files you'll need to create to implement a custom widget:

- ✓ **Java implementation file** - This is the file that implements the behavior of the widget. If you can instantiate the object from layout XML, you will also have to code a constructor that retrieves all the attribute values from the layout XML file.
- ✓ **XML definition file** - An XML file in res/values/ that defines the XML element used to instantiate your widget, and the attributes that it supports. Other applications will use this element and attributes in their in another in their layout XML.
- ✓ **Layout XML** [*optional*] - An optional XML file inside res/layout/ that describes the layout of your widget.

TextView android widget is the simplest; it can display text on the screen and are not editable.

Text View

```
<TextView
android:layout_width="fill_parent"
android:layout_height="wrap_content"
android:padding="3px"
android:text="User name"
/>
```

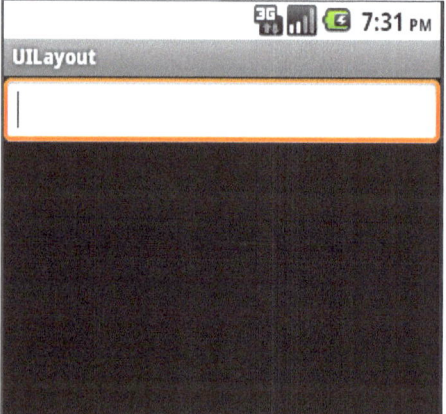

EditText widget

```
<EditText
    android:layout_width="fill_parent"
    android:layout_height="wrap_content"
/>
```

EditText Attribute

Attributes	Description
android:cursorVisible	Makes the cursor visible (the default) or invisible.
android:enabled	Specifies whether the TextView is enabled or not.
android:hint	Hint text to display when the text is empty.
android:maxLength	Set an input filter to constrain the text length to the specified number.
android:numeric	If set, specifies that this TextView has a numeric input method.
android:password	Whether the characters of the field are displayed as password dots instead of themselves.
android:phoneNumber	If set, specifies that this TextView has a phone number input method.
android:text	Text to display.
android:textColor	Text color.
android:textSize	Size of the text.
android:textStyle	Style (bold, italic, bold italic) for the text.
android:typeface	Typeface (normal, sans, serif, monospace) for the text.

Button

A button is a subclass of Textview

```
<Button
    android:layout_width="fill_parent"
    android:layout_height="wrap_content"
    android:text="Button"
    />
```

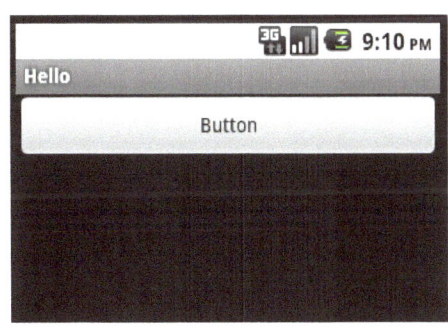

ImageView

Display a button with an image (instead of text) that can be pressed or clicked by the user. By default, an ImageButton looks like a regular Button, with the standard button background that changes color during different button states. The image on the surface of the button is defined either by the **android:src** attribute in the XML element

```
<ImageView
    android:layout_width="fill_parent"
    android:layout_height="wrap_content"
```

```
android:src="@drawable/icon"/>
```

Checkbox

A checkbox is a specific type of two-state button that can be either checked or unchecked. An example usage of a checkbox inside your activity would be the following:

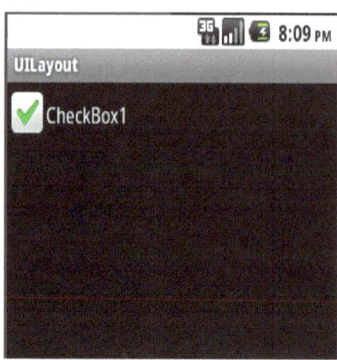

```
<CheckBox android:id="@+id/chkC1"
    android:layout_width="wrap_content"
    android:layout_height="wrap_content"
    android:text="CheckBox1" >
</CheckBox>
```

Radio Buttons

A radio button is a two-state button that can be either checked or unchecked. When the radio button is unchecked, the user can press or click it to check it. However, contrary to a CheckBox, a radio button cannot be unchecked by the user once checked.

Radio buttons are normally used together in a RadioGroup.
When several radio buttons live inside a radio group, checking one radio button uncheck all the others.

```
<RadioGroup
    android:orientation="vertical"
    android:layout_width="fill_parent"
    android:layout_height="fill_parent"
    >
    <RadioButton
        android:layout_width="wrap_content"
        android:layout_height="wrap_content"
        android:text="Radio Button 1" />
    <RadioButton
        android:layout_width="wrap_content"
        android:layout_height="wrap_content"
        android:text="Radio Button 2" />
    <RadioButton
        android:layout_width="wrap_content"
        android:layout_height="wrap_content"
        android:text="Radio Button 3" />
</RadioGroup>
```

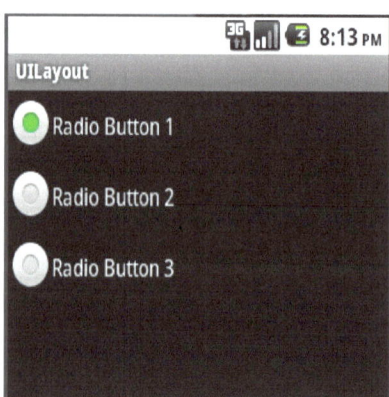

RadioGroup Attributes

Attribute	Description
android:checkedButton	The id of the child radio button that should be checked by default within this radio group.
android:orientation	Should the radio group be a column or a row? Use "horizontal" for a row, "vertical" for a column.

Listview

Listview allows you to show a List, you can do that in two ways ListView and ListActivity

```xml
<?xml version="1.0" encoding="utf-8"?>
<LinearLayout
xmlns:android="http://schemas.android.com/apk/res/android"
android:orientation="vertical"
android:layout_width="fill_parent"
android:layout_height="fill_parent" >

<ListView
 android:id="@android:id/list"
 android:layout_width="fill_parent"
 android:layout_height="fill_parent"
 android:drawSelectorOnTop="false" />

</LinearLayout>
```

You can create an XML resources file and load the data in the ListView from that XML file at runtime.

Spinner

```xml
<?xml version="1.0" encoding="utf-8"?>
<LinearLayout xmlns:android=http://schemas.android.com/apk/res/android
android:id="@+id/Linear"
android:layout_width="fill_parent"
android:layout_height="fill_parent"
android:orientation="vertical">

<TextView
android:id="@+id/select"
android:layout_width="fill_parent"
android:layout_height="wrap_content"
android:text="Select temperature"
android:textSize="8pt"
android:textStyle="bold"
>
</TextView>
```

```
<Spinner
android:id="@+id/spinner"
android:entries="@array/spinner1_list"
android:layout_width="fill_parent"
android:layout_height="wrap_content"
>
</Spinner>
</LinearLayout>
```

N.B to view the lists you must create a file containing the resource

```
<resources>
<string name="app_name">TemperatureConverter</string>
<string-array name="spinner1_list">
<item>Fahrenheit</item>
<item>Celsius</item>
<item>Kelvin</item>
</string-array>
</resources>
```

Gridview

A GridView is a ViewGroup that displays items in a two-dimensional, scrollable grid. The grid items are automatically inserted to the layout using a ListAdapter.

Chapter 4: Designing UI with Layout

FrameLayout

A FrameLayout is the simplest type of layout object. It's basically a blank space on your screen that you can later fill with a single object for example, a picture that you'll swap in and out.

All child elements of the FrameLayout are pinned to the top left corner of the screen; you cannot specify a different location for a child view.

Subsequent child views will simply be drawn over previous ones, partially or totally obscuring them (unless the newer object is transparent).

```xml
<?xml version="1.0" encoding="utf-8"?>
<FrameLayout xmlns:android="http://schemas.android.com/apk/res/android"
    android:orientation="vertical"
    android:layout_width="fill_parent"
    android:layout_height="fill_parent"
    >
<TextView
    android:layout_width="fill_parent"
    android:layout_height="wrap_content"
    android:text="@string/hello"
    />
</FrameLayout>
```

LinearLayout

LinearLayout aligns all children in a single direction — vertically or horizontally, depending on how you define the **orientation** attribute.

All children are stacked one after the other so a

Vertical list will only have one child per row, no matter how wide they are, and a Horizontal list will only be one row high (the height of the tallest child, plus padding)

A LinearLayout respects *margin*s between children and the *gravity* (right, center, or left alignment) of each child.

You may attribute a **weight** to children of a LinearLayout.
Weight gives an "importance" value to a view, and allows it to expand to fill any remaining space in the parent view.

```xml
<?xml version="1.0" encoding="utf-8"?>
<LinearLayout xmlns:android="http://schemas.android.com/apk/res/android"
    android:orientation="vertical"
    android:layout_width="fill_parent"
    android:layout_height="fill_parent"
    >
<TextView
    android:layout_width="fill_parent"
    android:layout_height="wrap_content"
    android:text="@string/hello"
    />
</LinearLayout>
```

android:orientation=*"vertical"* : LinearLayout will put widgets one after the other vertically (column)

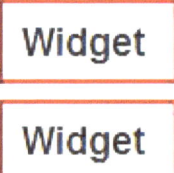

android:orientation=*"horizontal"* : LinearLayout will put widgets one after the other horizontally (online)

For all widgets inside a LinearLayout you must supply **android:layout_width** and **android:layout_height** properties.

Wrap_content : the widget should fill up its natural space
Fill_parent : the widget should fill up all available space in its enclosing container

TableLayout

TableLayout positions its children into **rows** and **columns**.
TableLayout containers do not display border lines.
The table will have as many columns as the row with the most cells.
A cell could be empty, but *cannot span columns*, as they can in HTML.
A *TableRow* object defines a single row in the table.
A row has zero or more cells; each cell is defined by any kind of other View.
A cell may also be a ViewGroup object. (For example, you can nest another TableLayout as a cell).

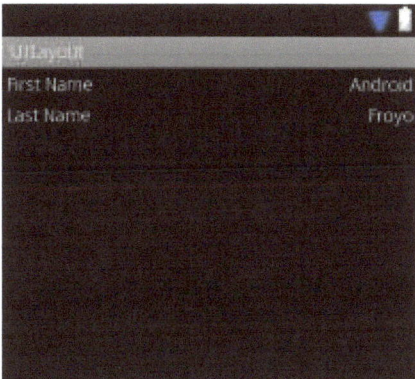

RelativeLayout

RelativeLayout lets child views specify their position relative to the parent view or to each other (specified by ID). So you can align two elements by right border, or make one below another, centered in the screen, centered left, and so on.

Elements are rendered in the order given, so if the first element is centered in the screen, other elements aligning themselves to that element will be aligned relative to screen center.

Also, because of this ordering, if using XML to specify this layout, the element that you will reference (in order to position other view objects) must be listed in the XML file before you refer to it from the other views via its reference ID.

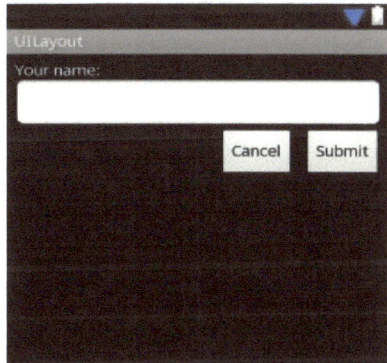

Adding background image to layout

1- Go to your Hello Android project created in *chapter 1*
2- Copy the background image in the **res/drawable** folder
3- Open the main.xml located in the **res/layout** folder and delete its content
4- Copy the following xml code

```xml
<?xml version="1.0" encoding="utf-8"?>

<LinearLayout
xmlns:android="http://schemas.android.com/apk/res/android"
        android:layout_width="fill_parent"
        android:layout_height="fill_parent"
        android:background="@drawable/bg"
        android:orientation="horizontal">

</LinearLayout>
```

android:background=*"@drawable/bg"* where bg is the name of your background image.

This attribute allows you to add a background image to your layout.

5- Run the application, and Hello android project will look like the following screenshot:

ScrollView Layout

When you have more data than what can be shown on a single screen use the **ScrollView** control.
It provides a sliding or scrolling access to the data. This way the user can only see part of your layout at one time, but the rest is available via scrolling.

ScrollView example

1- Create a new project and name it scrollview
2- Go in **res/layout** directory open **main.xml** file and delete its content
3- Add the following xml code in **main.xml**

```xml
<?xml version="1.0" encoding="utf-8"?>
<ScrollView xmlns:android="http://schemas.android.com/apk/res/android"
android:id="@+id/Scroll"
android:layout_width="fill_parent"
android:layout_height="fill_parent"
android:background="#2B2B2B"

<LinearLayout
android:id="@+id/Layout1"
android:layout_width="fill_parent"
android:layout_height="fill_parent"
android:orientation="vertical"
>
</LinearLayout>
</ScrollView>
```

4- Between <LinearLayout> and </LinearLayout>, add a horizontal LinearLayout containing your list, this time with a horizontal orientation

```xml
<LinearLayout
android:id="@+id/Layout2"
android:layout_width="fill_parent"
android:layout_height="fill_parent"
android:orientation="horizontal"
>

</LinearLayout>
```

5- Add a textview control within the horizontal linearlayout, ID is associated with them so that they can be referenced inside the source code of the activity.

Set the layout width and height attributes to **fill_parent** and **wrap_content** respectively

```xml
<TextView
```

```
android:id="@+id/textView1"
android:layout_width="fill_parent"
android:layout_height="wrap_content"
>
```

6- And then give to your TextView a text to be displayed to the screen, a size, font and color with the following attributes :

```
android:text="Cupcake"
android:textSize="50dip"
android:typeface="monospace"
android::textColor="#ffffff"
```

7- Now add a horizontal line after each textview element with the <view> object and give a background color to your view

```
<View
android:layout_width="fill_parent"
android:layout_height="2dip"
android:background="#161616"/>
```

The attribute **android:layout_height="2dip"** allows you to give a height of 2 dip to your line.

8- Create another **view** and **TextView** label

Here's the XML

```
<View
android:layout_width="fill_parent"
android:layout_height="2dip"
android:background="#161616"/>
<TextView
android:id="@+id/textView2"
android:layout_width="fill_parent"
android:layout_height="wrap_content"
android:text="Donut"
android:textSize="50dip"
android:typeface="monospace"
android:textColor="#FF6A00"/>
<View
android:layout_width="fill_parent"
android:layout_height="2dip"
android:background="#161616"/>
<TextView
android:id="@+id/textView3"
android:layout_width="fill_parent"
android:layout_height="wrap_content"
```

```
        android:text="Eclair"
        android:textSize="50dip"
        android:typeface="monospace"
        android:textColor="#ffffff"/>
    <View
        android:layout_width="fill_parent"
        android:layout_height="2dip"
        android:background="#161616"/>
    <TextView
        android:id="@+id/textView4"
        android:layout_width="fill_parent"
        android:layout_height="wrap_content"
        android:text="Froyo"
        android:textSize="50dip"
        android:typeface="monospace"
        android:textColor="#FF6A00"/>
    <View
        android:layout_width="fill_parent"
        android:layout_height="2dip"
        android:background="#161616"/>
    <TextView
        android:id="@+id/textView5"
        android:layout_width="fill_parent"
        android:layout_height="wrap_content"
        android:text="GingerBread"
        android:textSize="50dip"
        android:typeface="monospace"
        android:textColor="#ffffff"/>
    <View
        android:layout_width="fill_parent"
        android:layout_height="2dip"
        android:background="#161616"/>
    <TextView
        android:id="@+id/textView3"
        android:layout_width="fill_parent"
        android:layout_height="wrap_content"
        android:text="HoneyComb"
        android:textSize="50dip"
        android:typeface="monospace"
        android:textColor="#FF6A00"/>

    <View
        android:layout_width="fill_parent"
        android:layout_height="2dip"
        android:background="#161616"/>
    <TextView
        android:id="@+id/textView3"
        android:layout_width="fill_parent"
        android:layout_height="wrap_content"
        android:text="IceCream"
```

```
android:textSize="50dip"
android:typeface="monospace"
android:textColor="#ffffff"/>
```

The resulting layout should look something like this:

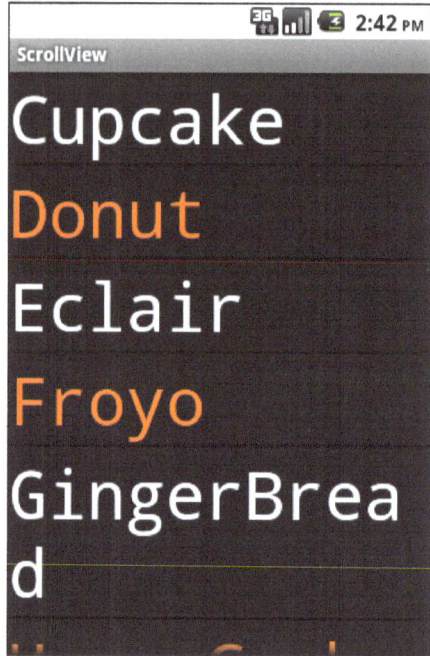

Case-study 1: create the UI of a form

The final output will be like below prototype

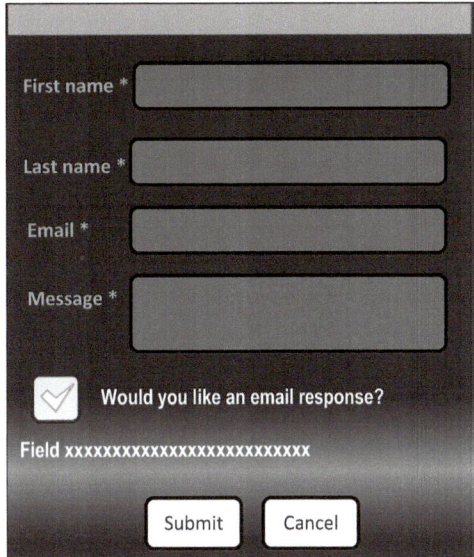

1- Create a new project and name it **Simple Form**
2- Create the layout of your form

Add the proprieties for your TableLayout, set the **android:layout_width** and **android:layout_height** to fill_parent".
Apply a margin at the top to leave a space between the layout and the title bar, and apply the **android:stretchColumns="1"** to specify column that should be stretched.

```xml
<?xml version="1.0" encoding="utf-8"?>
<TableLayout xmlns:android="http://schemas.android.com/apk/res/android"

    android:layout_width="fill_parent"
    android:layout_height="fill_parent"
    android:layout_marginTop="5sp"
    android:stretchColumns="1"
    >

</TableLayout>
```

3- Add the First Name label and the field receiving the user input within *TableRow*; remember that a *TableRow object* defines a single row in the table.

Add the tags <TableRow> and </TableRow>

```
<TableRow>
```

```
<TextView
    android:layout_width="wrap_content"
    android:layout_height="wrap_content"
    android:text="First name *"
    android:textColor="#FFFFFF"
    />
<EditText
    android:layout_width="fill_parent"
    android:layout_height="wrap_content"
    android:layout_margin="6dip"
    />
</TableRow>
```

4- Add the other fields (Last name, Email, Message) within <TableRow> ... </TableRow>
   tags

Note: The message field as presented earlier in the prototype can take several lines of text.
To do so, you will set the inputType attribute of the EditText control to TextMultiline and
specify the number of lines using the **android:lines** attribute.

```
<TableRow>

  <TextView
      android:layout_width="wrap_content"
      android:layout_height="wrap_content"
      android:text="Last name *"
      android:textColor="#FFFFFF"
      />
  <EditText
      android:layout_width="fill_parent"
      android:layout_height="wrap_content"
      android:layout_margin="6dip"
      />
</TableRow>

<TableRow>
  <TextView
      android:layout_width="wrap_content"
      android:layout_height="wrap_content"
      android:text="Email *"
      android:textColor="#FFFFFF"
      />
  <EditText
      android:layout_width="200px"
      android:layout_height="wrap_content"
      android:layout_margin="6dip"
      android:paddingRight="5dip"
      />
```

```
      </TableRow>

      <TableRow>
        <TextView
          android:layout_width="wrap_content"
          android:layout_height="wrap_content"
          android:text="Message"
          android:textColor="#FFFFFF"
          />

         <EditText
          android:layout_width="fill_parent"
          android:layout_height="wrap_content"
          android:layout_margin="6dip"
          android:paddingRight="5dip"
          android:inputType="textMultiLine"
          android:lines="3"
         />
</TableRow>
</TableRow>
```

5- Add a LinearLayout inside your TableLayout

```
<LinearLayout
     android:layout_width="wrap_content"
     android:layout_height="wrap_content">
</LinearLayout>
```

6- Add a CheckBox control inside your linearlayout

```
<CheckBox android:id="@+id/chkC1"
android:layout_width="wrap_content"
android:layout_height="wrap_content"
android:text="Would you like an email response?"
  >
  </CheckBox>
```

7- Add a Textview element below the checkbox control; put it in a LinearLayout.

```
<LinearLayout
     android:layout_width="wrap_content"
     android:layout_height="wrap_content"
     android:layout_marginTop="3sp">
   <TextView
     android:layout_width="wrap_content"
     android:layout_height="wrap_content"
     android:layout_margin="10dip"
     android:text="Fields marked with an * are required"
     android:textColor="#FFFFFF"
     />
```

```
        </LinearLayout>
```
8- Now add two Buttons controls "Submit" and "Cancel", to finish put them in a
   LinearLayout.

```
<LinearLayout
 android:orientation="horizontal"
 android:layout_width="wrap_content"
  android:layout_height="wrap_content"
  android:layout_marginTop="15sp"
  android:gravity="center"
  >
  <Button
  android:layout_width="wrap_content"
  android:layout_height="wrap_content"
  android:text="Submit"
  />
    <Button
  android:layout_width="wrap_content"
  android:layout_height="wrap_content"
  android:text="Cancel"
    />
</LinearLayout>
```

9- Save and **Run**

Chapter 5: Styles and Themes

A **style** is a collection of properties that specify the look and format for a **View** or window. A style can specify properties such as *height*, *padding*, *font color*, *font size*, *background color*, and much more.

A style is defined in an XML resource that is separate from the XML that specifies the layout.

Styles in Android share a similar philosophy to cascading style sheets in web design they allow you to separate the design from the content.

```
<TextView
    android:layout_width="fill_parent"
    android:layout_height="wrap_content"
    android:textColor="#00FF00"
    android:typeface="monospace"
    android:text="@string/hello" />
```

Will become this when working with style

```
<TextView
    style="@style/CodeFont"
    android:text="@string/hello" />
```

A **theme** is a style applied to an entire Activity or application, rather than an individual View (as in the example above). When a style is applied as a theme, every View in the Activity or application will apply each style property that it supports.

Defining Styles

To create a set of styles, save an XML file in the res/values/ directory of your project

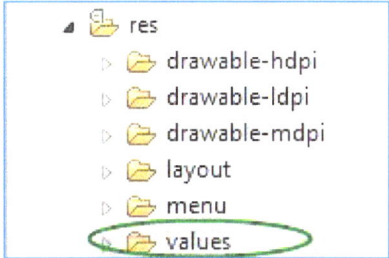

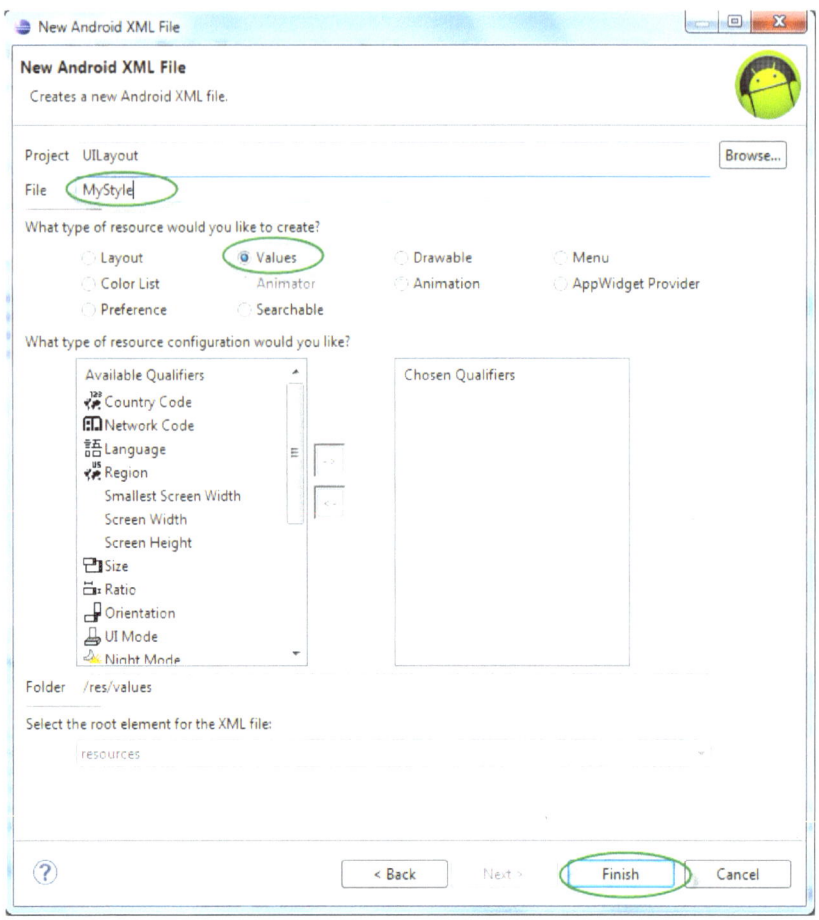

The file **MyStyle** is created in the value folder

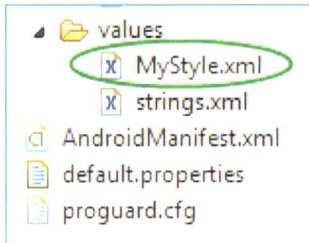

The root node of the XML file must be **<resources>**.

Double click on MyStyle.xml you will see the root node of the XML file like this

```
<?xml version="1.0" encoding="utf-8"?>
<resources>
...
</resources>
```

For each style you want to create, add a <style> element to the file with a name that uniquely identifies the style (this attribute is required).

```
<style name="NameOfTheStyle"
</style>
```

Add an <item> element for each property of that style, with a name that declares the style property and a value to go with it (this attribute is required).

```
<item name="android:layout_width">fill_parent</item>
```

The value for the <item> can be:

A keyword string

```
<item name="android:layout_width">fill_parent</item>
```

A hex color

```
<item name="android:textColor">#00FF00</item>
```

the style *"MyStyle.xml"* should look like this:

```
<?xml version="1.0" encoding="utf-8"?>
<resources>
    <style name=" FirstStyle " parent="@android:style/TextAppearance.Medium">
        <item name="android:layout_width">fill_parent</item>
        <item name="android:layout_height">wrap_content</item>
        <item name="android:textColor">#00FF00</item>
        <item name="android:typeface">monospace</item>
    </style>
</resources>
```

For example, you can apply the same *FirstStyle* style as a theme for an Activity and then all text inside that Activity will have green monospace font.

The parent attribute in the <style> element is optional and specifies the resource ID of another style from which this style should inherit properties.

Inheritance

The parent attribute in the <style> element lets you specify a style from which your style should inherit properties. You can use this to inherit properties from an existing style and then define only the properties that you want to change or add.
You can inherit from styles that you've created yourself or from styles that are built into the platform.

```
<style name="GreenText" parent="@android:style/TextAppearance">
        <item name="android:textColor">#00FF00</item>
</style>
```

If you want to inherit from styles that you've defined yourself, you *do not* have to use the *parent* attribute. Instead, just prefix the name of the style you want to inherit to the name of your new style, separated by a period.

```
<style name="CodeFont.Red">
   <item name="android:textColor">#FF0000</item>
 </style>
```

> **Note:** This technique for inheritance by chaining together names only works for styles defined by your own resources. You can't inherit Android built-in styles this way.

Style Properties

Now that you understand how a style is defined, you need to learn what kind of style properties defined by the <item> element is available.

You're probably familiar with some already, such as layout_width and textColor. Of course, there are many more style properties you can use.

Applying styles and themes to the UI

There are two ways to set a style:

1- To an individual View, by adding the *style* attribute to a View element in the XML for your layout.
2- Or, to an entire Activity or application, by adding the *android:theme* attribute to the <activity> or <application> element in the Android manifest.

When you apply a style to a single View in the layout, the properties defined by the style are applied only to that View.

If a style is applied to a ViewGroup, the child View elements will not inherit the style properties only the element to which you directly apply the style will apply its properties.

However, you *can* apply a style so that it applies to all View elements by applying the style as a theme.

To apply a style definition as a theme, you must apply the style to an Activity or application in the Android manifest.

Apply a style to a View

Here's how to set a style for a View in the XML layout:

```
<TextView
style="@style/FirstStyle"

android:id="@+id/select"
```

```
android:text="Select temperature">
</TextView>
```

Now this TextView has a custom style

Apply a theme to an Activity or application

To set a theme for all the activities of your application, open the *AndroidManifest.xml* file and edit the <application> tag to include the *android:theme* attribute with the style name.

1- Open the AndroidManifest.xml
2- Click **Application** tab at the bottom of the window
3- Click browse (theme)

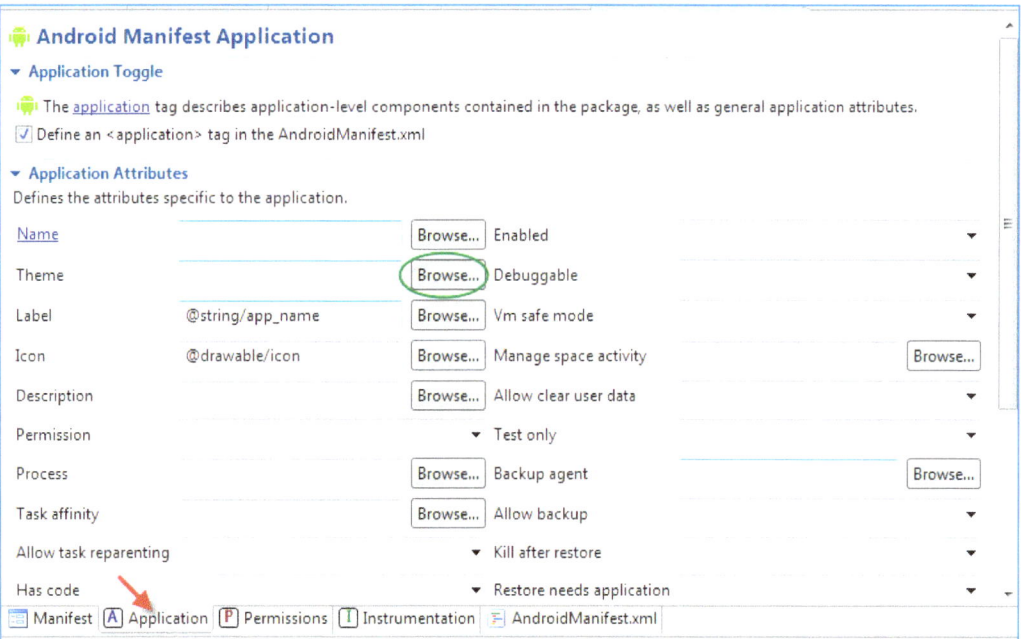

4- A new window appears (Resource Chooser)

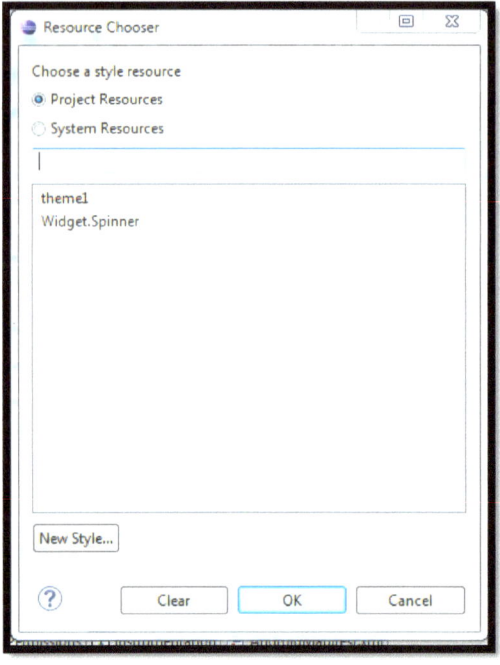

```
<application android:theme="@style/CustomTheme">
```

If you want a theme applied to just one Activity in your application, then add the *android:theme* attribute to the <activity> tag instead.

Just as Android provides other built-in resources, there are many pre-defined themes that you can use, to avoid writing them yourself.

You can access these built-in resources while opening the *AndroidManifest.xml* file and click <application> tag

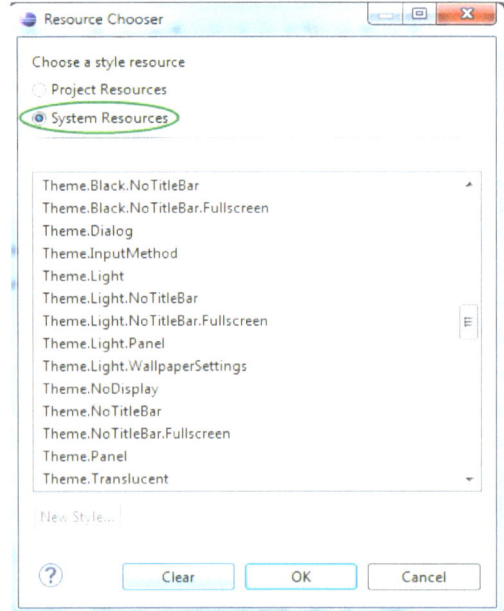

For example, you can use the *Dialog* theme and make your Activity appear like a dialog box

```
<activity android:theme="@android:style/Theme.Dialog">
```

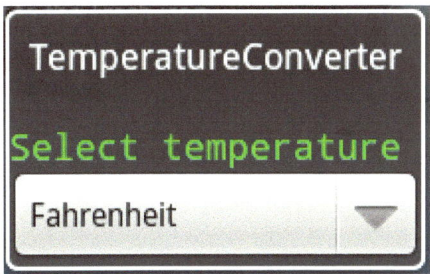

You can use the Light theme:

```
<activity android:theme="@android:style/Theme.Light">
```

If you like a theme, but want to tweak it, just adds the theme as the *parent* of your custom theme. For example, you can modify the traditional light theme to use your own color like this:

```
<color name="custom_theme_color">#b0b0ff</color>
<style name="CustomTheme" parent="android:Theme.Light">
    <Item name="android:windowBackground">@color/custom_theme_color</item>
    <Item name="android:colorBackground">@color/custom_theme_color</item>
</style>
```

Select a theme based on platform version

Newer versions of Android have additional themes available to applications, and you might want to use these while running on those platforms while still being compatible with older versions. You can accomplish this through a custom theme that uses resource selection to switch between different parent themes, based on the platform version.

For example, here is the declaration for a custom theme which is simply the standard platforms default light theme. It would go in an XML file under res/values (typically res/values/styles.xml):

```
<style name="LightThemeSelector" parent="android:Theme.Light">
    ...
</style>
```

To have this theme use the newer holographic theme when the application is running on Android 3.0 (API Level 11) or higher, you can place an alternative declaration for the theme in an XML file in res/values-v11, but make the parent theme the holographic theme:

```
<style name="LightThemeSelector" parent="android:Theme.Holo.Light">
    ...
</style>
```

Chapter 6: Application Resources

You should always externalize application resources such as images and strings from your code, so that you can maintain them independently. You should also provide alternative resources for specific device configurations, by grouping them in specially-named resource directories. At runtime, Android uses the appropriate resource based on the current configuration. For example, you might want to provide a different UI layout depending on the screen size or different strings depending on the language setting.

Once you externalize your application resources, you can access them using resource IDs that are generated in your project's *R* class.

Grouping Resource Types

You should place each type of resource in a specific subdirectory of your project's res/ directory. For example, here's the file hierarchy for a simple project

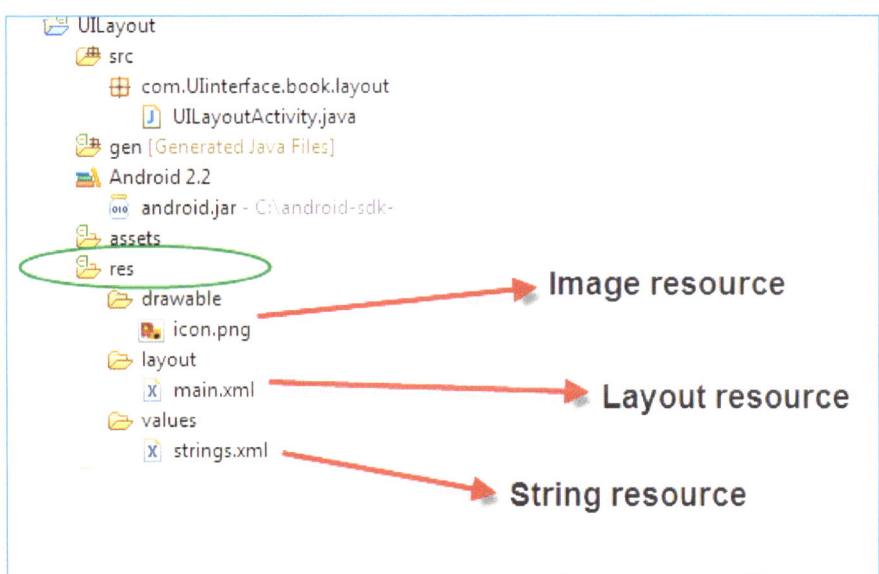

As you can see in this image, the res/ directory contains all the resources (in subdirectories): an image resource, a layout resource, and a string resource file. The resource directory names are important and are described in **Appendix A**

The resources that you save in the subdirectories defined in **Appendix A** are your "default" resources. That is, these resources define the default design and content for your application. However, different types of Android-powered devices might call for different types of resources.

For example, if a device has a larger than normal screen, then you should provide different layout resources that takes advantage of the extra screen space. Or, if a device has a different language setting, then you should provide different string resources that translate the text in your user interface. To provide these different resources for different device configurations, you need to provide alternative resources, in addition to your default resources.

Resource Types

Here's a brief summary of each resource type:

Resource Types	Description
Animation Resources	Define pre-determined animations. Tween animations are saved in res/anim/ and accessed from the R.anim class. Frame animations are saved in res/drawable/ and accessed from the R.drawable class.
Color State List Resource	Define a color resource that changes based on the View state. Saved in res/color/ and accessed from the R.color class.
Drawable Resources	Define various graphics with bitmaps or XML. Saved in res/drawable/ and accessed from the R.drawable class.
Layout Resource	Define the layout for your application UI. Saved in res/layout/ and accessed from the R.layout class.
Menu Resource	Define the contents of your application menus. Saved in res/menu/ and accessed from the R.menu class.
String Resources	Define strings, string arrays, and plurals (and include string formatting and styling). Saved in res/values/ and accessed from the R.string, R.array, and R.plurals classes.
Style Resource	Define the look and format for UI elements. Saved in res/values/ and accessed from the R.style class.

More Resource Types

Define values such as booleans, integers, dimensions, colors, and other arrays.
Saved in res/values/ but each accessed from unique R sub-classes (such as R.bool, R.integer, R.dimen, etc.).

Providing Alternative Resources

Support specific device configurations.

To specify configuration-specific alternatives for a set of resources:

Create a new directory in res/ named in the form **<resources_name>-<config_qualifier>**.
<resources_name> is the directory name of the corresponding default resources.
<qualifier> is a name that specifies an individual configuration for which these resources are to be used *(see appendix A)*.

You can append more than one *<qualifier>*. Separate each one with a dash.
Save the respective alternative resources in this new directory. The resource files must be named exactly the same as the default resource files.

For example, here are some default and alternative resources:

res/

　　drawable/

　　　　icon.png

　　　　background.png

　　drawable-hdpi/

　　　　icon.png

　　　　background.png

The *hdpi* qualifier indicates that the resources in that directory are for devices with a high-density screen. The images in each of these drawable directories are sized for a specific screen density, but the filenames are exactly the same. This way, the resource ID that you use to reference the *icon.png* or *background.png* image is always the same, but Android selects the version of each resource that best matches the current device, by comparing the device configuration information with the qualifiers in the alternative resource directory name.

You can add multiple qualifiers to one directory name, by separating each qualifier with a dash.

Appendix A lists the valid configuration qualifiers, in order of precedence if you use multiple qualifiers for one resource directory, they must be added to the directory name in the order they are listed.

Qualifier name rules

Here are some rules about using configuration qualifier names:

You can specify multiple qualifiers for a single set of resources, separated by dashes. For example, drawable-en-rUS-land applies to US-English devices in landscape orientation.

The qualifiers must be in the order listed in **Appendix A**. For example:
Wrong: drawable-hdpi-port/
Correct: drawable-port-hdpi/

Alternative resource directories cannot be nested. For example, you cannot have res/drawable/drawable-en/.

Values are case-insensitive. The resource compiler converts directory names to lower case before processing to avoid problems on case-insensitive file systems. Any capitalization in the names is only to benefit readability.

Only one value for each qualifier type is supported.

For example, if you want to use the same drawable files for Spain and France, you *cannot* have a directory named drawable-rES-rFR/. Instead you need two resource directories, such as drawable-rES/ and drawable-rFR/, which contain the appropriate files. However, you are not required to actually duplicate the same files in both locations. Instead, you can create an alias to a resource.

Creating alias resources

When you have a resource that you'd like to use for more than one device configuration (but do not want to provide as a default resource), you do not need to put the same resource in more than one alternative resource directory. Instead, you can (in some cases) create an alternative resource that acts as an alias for a resource saved in your default resource directory.

Drawable

To create an alias to an existing drawable, use the <bitmap> element. For example:

```xml
<?xml version="1.0" encoding="utf-8"?>
<bitmap xmlns:android="http://schemas.android.com/apk/res/android"
 android:src="@drawable/icon_ca"/>
```

If you save this file as *icon.xml* (in an alternative resource directory, such as res/drawable-en-rCA/), it is compiled into a resource that you can reference as R.drawable.icon, but is actually an alias for the R.drawable.icon_ca resource (which is saved in res/drawable/).

Layout

To create an alias to an existing layout, use the <include> element, wrapped in a <merge>

For example:

```xml
<?xml version="1.0" encoding="utf-8"?>
<merge>
    <include layout="@layout/main_ltr"/>
</merge>
```

Strings and other simple values

To create an alias to an existing string, simply use the resource ID of the desired string as the value for the new string. For example:

```xml
<?xml version="1.0" encoding="utf-8"?>
<resources>
    <string name="hello">Hello</string>
    <string name="hi">@string/hello</string>
</resources>
```

Color

```xml
<?xml version="1.0" encoding="utf-8"?>
<resources>
    <color name="yellow">#f00</color>
```

```
    <color name="highlight">@color/red</color>
  </resources>
```

Providing the Best Device Compatibility with Resources

- Always provide default resources for each type of resource that your application uses.

- If you provide different layout resources based on the screen orientation, you should pick one orientation as your default.

- Instead of providing layout resources in layout-land/ for landscape and layout-port/ for portrait, leave one as the default, such as layout/ for landscape and layout-port/ for portrait.

- Providing default resources is important not only because your application might run on a configuration you had not anticipated, but also because new versions of Android sometimes add configuration qualifiers that older versions do not support.

- Always provide default resources for the resources your application needs to perform properly

Accessing Resources from XML

You can define values for some XML attributes and elements using a reference to an existing resource. You will often do this when creating layout files, if you add a Button to your layout, you should use a *string resource* for the button text:

```
<Button
    android:layout_width="fill_parent"
    android:layout_height="wrap_content"
    android:text="@string/submit" />
```

Syntax

Here is the syntax to reference a resource in an XML resource:

@ [*<package_name>*:]*<resource_type>*/*<resource_name>*

<package_name> is the name of the package in which the resource is located (not required when referencing resources from the same package)
<resource_type> is the *R* subclass for the resource type
<resource_name> is either the resource filename without the extension or the android:name attribute value in the XML element (for simple values).

For example, if you have the following resource file that includes a color resource and a string resource:

```
<?xml version="1.0" encoding="utf-8"?>
<resources>
    <color name="opaque_red">#f00</color>
    <string name="hello">Hello!</string>
</resources>
```

You can use these resources in the following layout file to set the text color and text string:

```
<?xml version="1.0" encoding="utf-8"?>
<EditText xmlns:android="http://schemas.android.com/apk/res/android"
    android:layout_width="fill_parent"
    android:layout_height="fill_parent"
    android:textColor="@color/opaque_red"
    android:text="@string/hello" />
```

In this case you don't need to specify the **package name** in the resource reference because the resources are from your own package.

To reference a system resource, you would need to include the package name.

Example:

```
<?xml version="1.0" encoding="utf-8"?>
<EditText xmlns:android="http://schemas.android.com/apk/res/android"
    android:layout_width="fill_parent"
    android:layout_height="fill_parent"
    android:textColor="@android:color/secondary_text_dark"
    android:text="@string/hello" />
```

Note: You should use string resources at all times, so that your application can be localized for other languages.

You can even use resources in XML to create aliases.

For example, you can create a drawable resource that is an alias for another drawable resource:

```xml
<?xml version="1.0" encoding="utf-8"?>
  <bitmap xmlns:android="http://schemas.android.com/apk/res/android"
    android:src="@drawable/other_drawable" />
```

This sounds redundant, but can be very useful when using alternative resource.

Referencing style attributes

A style attribute resource allows you to reference the value of an attribute in the currently-applied theme. Referencing a style attribute allows you to customize the look of UI elements by styling them to match standard variations supplied by the current theme, instead of supplying a hard-coded value. Referencing a style attribute essentially says, "Use the style that is defined by this attribute, in the current theme."

To reference a style attribute, the name syntax is almost identical to the normal resource format, but instead of the at-symbol (@), use a question-mark (?), and the resource type portion is optional. For instance:

```
?[<package_name>:][<resource_type>/]<resource_name>
```

For example, here's how you can reference an attribute to set the text color to match the "primary" text color of the system theme:

```xml
<EditText id="text"
    android:layout_width="fill_parent"
    android:layout_height="wrap_content"
    android:textColor="?android:textColorSecondary"
    android:text="@string/hello_world" />
```

Here, the *android:textColor* attribute specifies the name of a style attribute in the current theme. Android now uses the value applied to the *android:textColorSeconday* style attribute as the value for *android:textColor* in this widget. Because the system resource tool knows that an attribute resource is expected in this context, you do not need to explicitly state the type (which would be ?android:attr/textColorSecondary) you can exclude the *attr* type.

Localization

Android will run on many devices in many regions. To reach the most users, your application should handle text, audio files, numbers, currency, and graphics in ways appropriate to the locales where your application will be used.

This chapter describes best practices for localizing Android applications. The principles apply whether you are developing your application using ADT with Eclipse, Ant-based tools, or any other IDE.

It is good practice to use the Android resource framework to separate the localized aspects of your application as much as possible from the core Java functionality:

Plan the localization

The first step in localizing an application is to plan how the application will render differently in different locales. In this application, the default locale will be the United Kingdom add some locale-specific information for France, Canada, Spain, and the United States.

Region / Language	United Kingdom	France	Canada	Spain	United States
English	British English text; British flag *(default)*	-	British English text; Canadian flag	-	British English text; U.S. flag
French	-	French text for app_name, text_a and text_b; French flag	French text for app_name, text_a and text_b; Canadian flag	-	-
Spanish	-	-	-	Spanish text for text_a and text_b; Spain flag	-

As shown in the table above, the plan calls for four flag icons in addition to the British flag that is already in the **res/drawable/** folder. It also calls text strings other than the text that is in **res/values/strings.xml**.

The following table shows where the needed text strings and flag icons will go, and specifies which ones will be loaded for which locales.

Locale Code	Language / Country	Location of strings.xml	Location of flag.png
Default	English / United Kingdom	res/values/	res/drawable/
fr-rFR	French / France	res/values-fr/	res/drawable-fr-rFR/
fr-rCA	French / Canada	res/values-fr/	res/drawable-fr-rCA/
en-rCA	English / Canada	*(res/values/)*	res/drawable-en-rCA/
es-rVE	Spanish/Venezuela	/res/values-es	/res/values/es-rVE
en-rUS	English / United States	*(res/values/)*	res/drawable-en-rUS/

Tip: A folder qualifer cannot specify a region without a language. Having a folder named res/drawable-rCA/, for example, will prevent the application from compiling

Localize the Applications

Localize the Strings

The application requires two more *strings.xml* files, one each for French, and Spanish. To create these resource files within Eclipse

1- Create a new project and name it localization

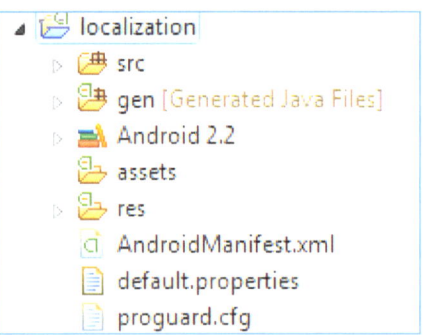

2- Right click in localization select **NEW > OTHERS > Android XML File** to open the New Android XML File wizard

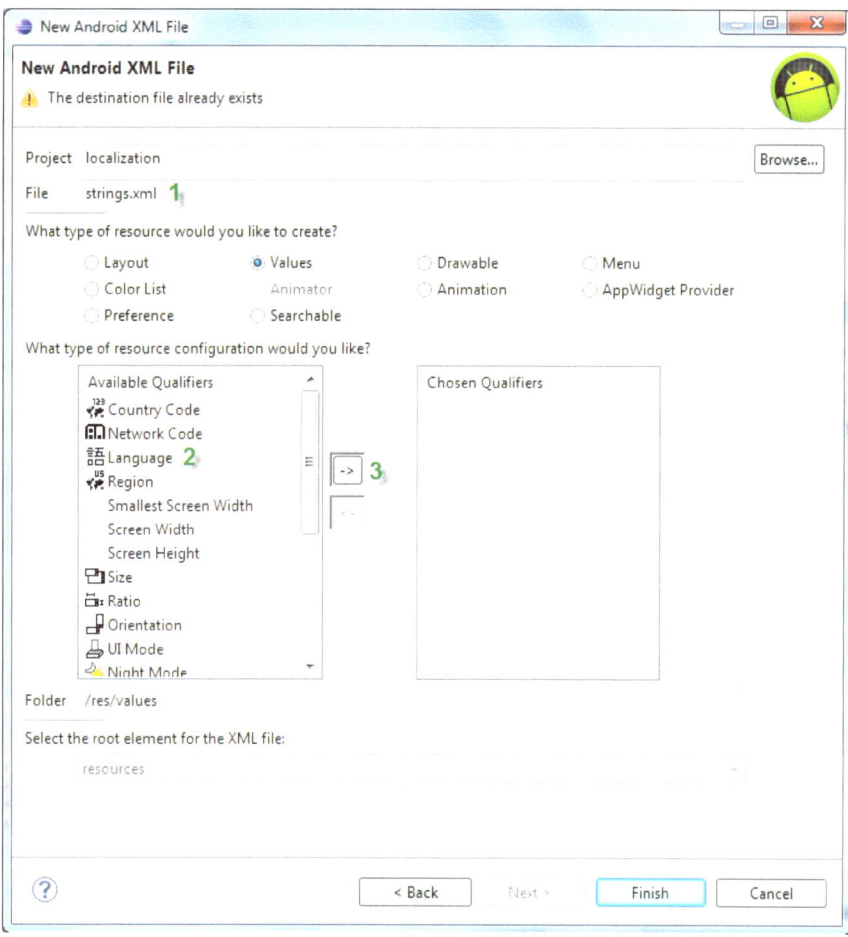

Specify the file name **1** and the type of resource **2** and click the arrow **3**

3- Type **fr** in the Language box and click Finish

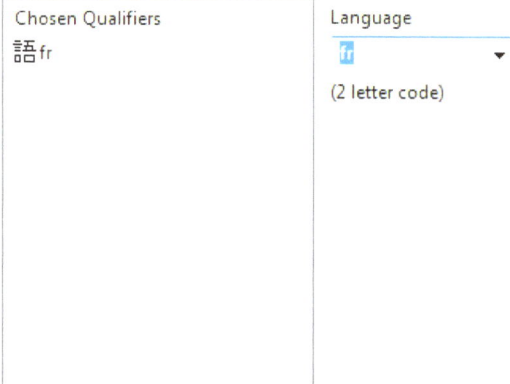

A new file, res/values-fr/strings.xml appears among the project files.

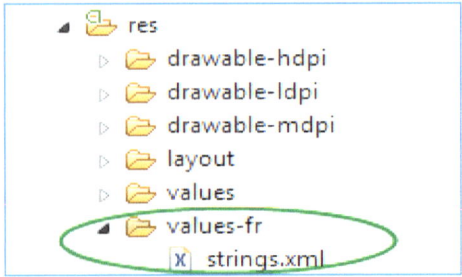

4- Repeat the steps for other language codes

Double click the **strings.xml** to add localized text to the new files as follows

```xml
<?xml version="1.0" encoding="utf-8"?>
<resources>
    <string name="app_name">bonjour, Android </string>
    <string name="text_a">francais francais francais</string>
    <string name="text_b">francais francais francais</string>
</resources>
```

Localize the Images

To locate your images you have to put icons in the images

Example res/drawable-fr-rFR/ for French

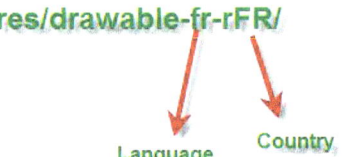

Default Resources

Why Default Resources Are Important

Whenever the application runs in a locale for which you have not provided locale-specific text, Android will load the default strings from res/values/strings.xml. If this default file is absent, or if it is missing a string that your application needs, then your application will not run and will show an error. The example below illustrates what can happen when the default text file is incomplete.

How to Create Default Resources

Put the application's default text in a file with the following location and name:

res/values/strings.xml (required directory)

The text strings in res/values/strings.xml should use the default language, which is the language that you expect most of your application's users to speak.

The default resource set must also include any default drawables and layouts, and can include other types of resources such as animations.

res/drawable/	required directory holding at least one graphic file, for the application's icon in the Market
res/layout/	required directory holding an XML file that defines the default layout
res/anim/	required if you have any res/anim-<*qualifiers*> folders
res/xml/	required if you have any res/xml-<*qualifiers*> folders
res/raw/	required if you have any res/raw-<*qualifiers*> folders

Tip: In your code, examine each reference to an Android resource. Make sure that a default resource is defined for each one. Also make sure that the default string file is complete: A localized string file can contain a subset of the strings, but the default string file must contain them all.

Chapter 7: Android Menu

Menus are an important part of an activity's user interface, which provide users a familiar way to perform actions. Android offers a simple framework for you to add standard menus to your application.

Beginning with Android 3.0 (API level 11), Android-powered devices are no longer required to provide a dedicated *Menu* button. With this change, Android apps should migrate away from a dependence on the traditional 6-item menu panel and instead provide an action bar to present common user actions.

Although the design and user experience for some menu items have changed, the semantics to define a set of actions and options is still based on the **Menu** APIs. In this chapter you will show you how to create the three fundamental types of menus or action presentations on all versions of Android:

Options Menu and action bar

The **options_menu** is the primary collection of menu items for an activity. It's where you should place actions that have a global impact on the app, such as "Search," "Compose email," and "Settings."

If you're developing for Android 2.3 or lower, users can reveal the options menu panel by pressing the *Menu* button.

On Android 3.0 and higher, items from the options menu are presented by the **action bar** as a combination of on-screen action items and overflow options. Beginning with Android 3.0, the *Menu* button is deprecated (some devices don't have one), so you should migrate toward using the action bar to provide access to actions and other options.

Context Menu and contextual action mode

A context menu is a **floating menu** that appears when the user performs a long-click on an element. It provides actions that affect the selected content or context frame.

When developing for Android 3.0 and higher, you should instead use the **contextual action mode** to enable actions on selected content. This mode displays action items that affect the selected content in a bar at the top of the screen and allows the user to select multiple items.

A popup menu displays a list of items in a vertical list that's anchored to the view that invoked the menu. It's good for providing an overflow of actions that relate to specific content or to provide options for a second part of a command. Actions in a popup menu should not directly affect the corresponding content—that's what contextual actions are for. Rather, the popup menu is for extended actions that relate to regions of content in your activity.

Creating a Menu Resources

Instead of instantiating a Menu in your application code, you should define a menu and all its items in an XML menu resource, then inflate the menu resource (load it as a programmable object) in your application code. Using a menu resource to define your menu is a good practice because it separates the content for the menu from your application code. It's also easier to visualize the structure and content of a menu in XML.

To create a menu resource, create an XML file inside your project res/menu/ directory and build the menu with the following elements:

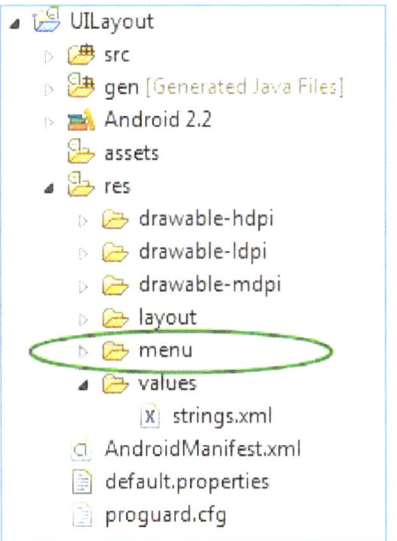

```xml
<?xml version="1.0" encoding="utf-8"?>

<menu xmlns:android="http://schemas.android.com/apk/res/android">

    <item android:id="@+id/menu1"
        android:icon="@drawable/ic_ic1"
        android:title="item1" />

    <item android:id="@+id/menu2"
        android:title="item2" />
```

```
<item android:id="@+id/menu3"
        android:title="item3" />
<item android:id="@+id/menu4"
        android:title="item4" />

</menu>
```

You must add icons and menu item in the **res/values**

```
<resources>

<string name="app_name">Menu</string>

    <string name="item1">Item 1</string>
    <string name="item2">Item 2</string>
    <string name="item3">Item 3</string>
    <string name="item4">Item 4</string>

</resources>
```

This example defines a menu with two items. Each item includes the attributes:

android:id

A resource ID that's unique to the item, which allows the application, can recognize the item when the user selects it.

android:icon

A reference to a drawable to use as the item's icon

android:title

A reference to a string to use as the item's title

Programming guide: From your application code, you can inflate a menu resource (convert the XML resource into a programmable object)

Creating an Options Menu

The options menu is where you should include actions and other options that are relevant to the current activity context, such as "Search," "Compose email," and "Settings."

Where the items in your options menu appear on the screen depends on the version for which you've developed your application:

o   If you've developed your application for **Android 2.3.x (API level 10) or lower**, the contents of your options menu appear at the bottom of the screen when the user presses the *Menu* button, as shown in figure 1. When opened, the first visible portion is the icon menu, which holds up to six menu items. If your menu includes more than six items, Android places the sixth item and the rest into the overflow menu, which the user can open by selecting **More**.

o If you've developed your application for **Android 3.0 (API level 11) and higher**, items from the options menu are available in the **action bar**. By default, the system places all items in the action overflow, which the user can reveal with the action overflow icon on the right side of the action bar (or by pressing the device *Menu* button, if available). To enable quick access to important actions, you can promote a few items to appear in the action bar by adding *android:showAsAction="ifRoom"* to the corresponding **<item>** elements

Note: Even if you're not developing for Android 3.0 or higher, you can build your own action bar layout for a similar effect.

Creating contextual menus

A contextual menu offers actions that affect a specific item or context frame in the UI. You can provide a context menu for any view, but they are most often used for items in a **ListView, GridView,** or other view collections in which the user can perform direct actions on each item.

There are two ways to provide contextual actions:

1- In a **floating context menu**. A menu appears as a floating list of menu items (similar to a dialog) when the user performs a long-click (press and hold) on a view that declares support for a context menu. Users can perform a contextual action on one item at a time.

2- In the **contextual action mode**. This mode is a system implementation of **ActionMode** that displays a *contextual action bar* at the top of the screen with action items that affect the selected item(s). When this mode is active, users can perform an action on multiple items at once (if your app allows it).

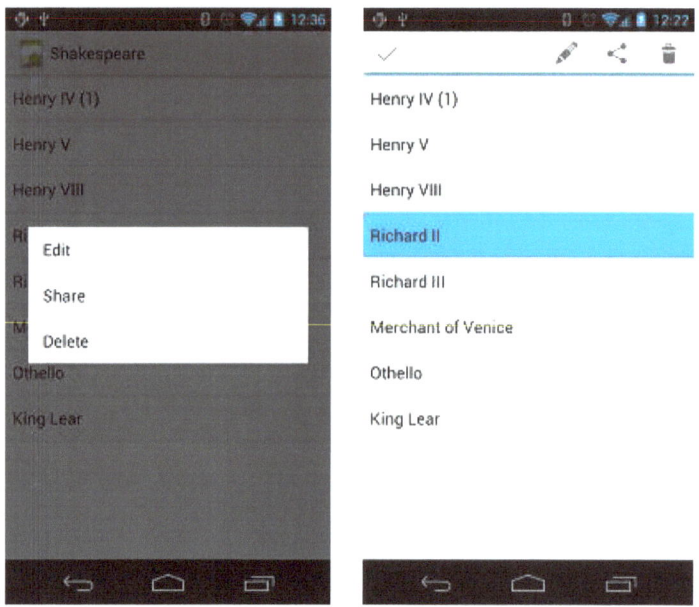

> Note: The contextual action mode is available on Android 3.0 (API level 11) and higher and is the preferred technique for displaying contextual actions when available. If your app supports versions lower than 3.0 then you should fall back to a floating context menu on those devices.

Creating a Popup Menu

A **PopupMenu** is a modal menu anchored to a **View**. It appears below the anchor view if there is room or above the view otherwise. It's useful for:

1- Providing an overflow-style menu for actions that *relate to* specific content (such as Gmail's email headers, shown in figure 4).

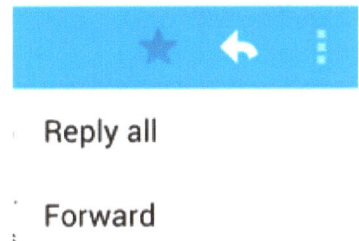

Reply all

Forward

Example

The final output will be like below prototype

1- Create a new project and name it Android Menu
2- Go in **res/layout** directory and open main.xml file
3- Add the following xml code in main.xml

```xml
<?xml version="1.0" encoding="utf-8"?>
<LinearLayout
xmlns:android="http://schemas.android.com/apk/res/android"
    android:orientation="vertical"
    android:layout_width="fill_parent"
    android:layout_height="fill_parent"
    android:background="#4a6c9b"
    >
<TextView
    android:layout_width="fill_parent"
    android:layout_height="wrap_content"
    android:paddingTop="50sp"
    android:gravity="center"
```

```
      android:text="@string/hello"
       android:textSize="30sp"
     />
</LinearLayout>
```

4- Add your menu Images in **res/drawable** folder

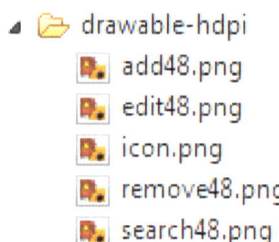

5- Go in **res/values** directory and open strings.xml, add the following XML code in strings.xml

```
<?xml version="1.0" encoding="utf-8"?>
<resources>
        <string name="hello">Member Login</string>
        <string name="app_name">Android Menu</string>
        <string name="menu_add">Add</string>
        <string name="menu_search">Search</string>
        <string name="menu_edit">Edit</string>
        <string name="menu_remove">Remove</string>
</resources>
```

6- Create the menu resource

Create an XML file inside your project's res/menu/ directory and build the menu with the following elements:

```
<?xml version="1.0" encoding="utf-8"?>
<menu
  xmlns:android="http://schemas.android.com/apk/res/android">
    <item android:id="@+id/menuItem1"
       android:icon="@drawable/add48"
       android:title="@string/menu_add"/>

    <item android:id="@+id/menuItem2"
       android:icon="@drawable/search48"
       android:title="@string/menu_search"/>

    <item android:id="@+id/menuItem3"
       android:icon="@drawable/edit48"
       android:title="@string/menu_edit"/>
```

```
    <item android:id="@+id/menuItem4"
        android:icon="@drawable/remove48"
        android:title="@string/menu_remove"/>

</menu>
```

7- **Run** the project, click the menu button in the emulator

Result should look like this:

Chapter 8: Merging Layout

Android allow merging two separates Layout into a single XML file.

The <merge /> tag was created for the purpose of optimizing Android layouts by reducing the number of levels in view trees.

For instance the <merge /> works perfectly when combined with the <include /> tag. You can also use <merge /> when you create a custom composite view. Here is the XML used to display this custom view on top of an image:

The final output will be like below prototype:

1- Create a new project and name it MergeLayout

This project contain a Header, a main container and a footer

Designing the header

You will design a background image for the header

Header background - tet.9.png (23px X 60 px)

Create a style for our project (see style chapter)

2- Go in **res/values** and create the MyStyle.xml file and copy the following code in your new created file

```xml
<?xml version="1.0" encoding="utf-8"?>
<resources>
    <style name="Header">
        <item name="android:Layout_width">fill_parent</item>
        <item name="android:Layout_height">50dp</item>
        <item name="android:orientation">horizontal</item>
        <item name="android:background">@drawable/tet</item>
    </style>

    <style name="Body">
        <item name="android:Layout_gravity">center_vertical</item>
        <item name="android:Layout_width">wrap_content</item>
        <item name="android:Layout_height">wrap_content</item>
        <item name="android:gravity">center_horizontal</item>
        <item name="android:drawablePadding">2dp</item>
        <item name="android:textSize">16dp</item>
        <item name="android:textStyle">bold</item>
        <item name="android:textColor">#ff29549f</item>
        <item name="android:background">@null</item>
    </style>

    <style name="Footer">
        <item name="android:Layout_width">fill_parent</item>
        <item name="android:Layout_height">40dp</item>
        <item name="android:orientation">horizontal</item>
        <item name="android:background">#dedede</item>
    </style>
</resources>
```

Create the header.xml

3- Go in **res/layout** directory create the header.xml and copy the following code

```xml
<LinearLayout
xmlns:android="http://schemas.android.com/apk/res/android"
    style="@style/Header" >
</LinearLayout>
```

Design the login form (Body)

4- Create a vertical LinearLayout, add the properties **android:layout_width** and **android:layout_height** to fill_parent, set the properties android:layout_weight="1", backgound color #040521, paddingTop="20sp"

```
<LinearLayout
xmlns:android="http://schemas.android.com/apk/res/android"
    android:orientation="vertical"
    android:layout_width="fill_parent"
    android:layout_height="fill_parent"
    android:layout_weight="1"
    android:background="#040521"
    android:paddingTop="20sp"
 >


</LinearLayout>
```

5- Between **<LinearLayout>** and **</LinearLayout>** add a Textview element

```
<TextView
    android:layout_width="wrap_content"
    android:layout_height="wrap_content"
    android:text="User Name"
    android:textColor="#FFFFFF"
    android:paddingTop="10dp" />
```

6- Add an EditText with **android:layout_width to** fill_parent and **android:layout_height** to wrap_content and a marging of 6 dip

```
<EditText
    android:layout_width="fill_parent"
    android:layout_height="wrap_content"
    android:layout_margin="6dip"
    />
```

7- Add another TextView elements with the properties **android:layout_width** and **android:layout_height** to wrap_content, and set the android text to "Password".

```
<TextView
    android:layout_width="wrap_content"
    android:layout_height="wrap_content"
    android:text="Password"
    android:textColor="#FFFFFF"/>
```

8- Add an EditText with **android:layout_width to** fill_parent and **android:layout_height** to wrap_content and a marging of 6 dip

```
<EditText
 android:layout_width="fill_parent"
 android:layout_height="wrap_content"
 android:layout_margin="6dip"/>
```

9- Then add another TextView elements with the properties **android:layout_width** and **android:layout_height** to wrap_content.

```
<TextView
    android:layout_width="wrap_content"
    android:layout_height="wrap_content"
    android:text="Forgot your username or password?"
    android:textColor="#FFFFFF"
    android:paddingTop="10dp"/>
```

Design the footer

You will create two buttons (submit, cancel) in the footer section

10- Go in **res/layout** directory create the footer.xml

11- Create a vertical Linearlayout, add the properties **android:layout_width** and **android:layout_height** to fill_parent and **android:layout_height** to wrap_content,a padding top of 6sp, put the layout at the bottom with the attribute android:gravity="*bottom*",android:layout_alignParentBottom="*true*" and a backgound color.

```
<LinearLayout
xmlns:android="http://schemas.android.com/apk/res/android"
  android:layout_width="fill_parent"
  android:layout_height="wrap_content"
  android:paddingTop="6sp"
  android:gravity="bottom"
  android:layout_alignParentBottom="true"
  android:background="#DFDFDF"
  >
</LinearLayout>
```

12- Between **<LinearLayout>** and **</LinearLayout>** add 2 buttons elements with the propeties **android:layout_width** to *wrap_content* and **android:layout_height** to 50px

```
<Button
    android:layout_width="wrap_content"
    android:layout_height="50px"
    android:id="@+id/submit"
    android:text="Submit"
    android:layout_alignParentBottom="true"
    android:layout_weight="1" />
  <Button
    android:layout_width="wrap_content"
    android:layout_height="50px"
    android:id="@+id/cancel"
    android:text="Cancel"
    android:layout_alignParentBottom="true"
    android:layout_weight="1" />
```

Now merge all layout together

13- Create a new XML file name it **merge_all.xml**

Create a vertical LinearLayout, add the propeties **android:layout_width** and **android:layout_height** to fill_parent, and merge all the layouts with the **<include>** tag

```
<LinearLayout
xmlns:android="http://schemas.android.com/apk/res/android"

    android:orientation="vertical"
    android:layout_width="fill_parent"
    android:layout_height="fill_parent">

        <!-- Include header -->

      <include layout="@layout/header"/>

        <!--  Include Body -->
        <include layout="@layout/body"/>

        <!--  Include Footer -->
        <include layout="@layout/footer"/>

    </LinearLayout>
```

14- Run the project

The final result should looks like this :

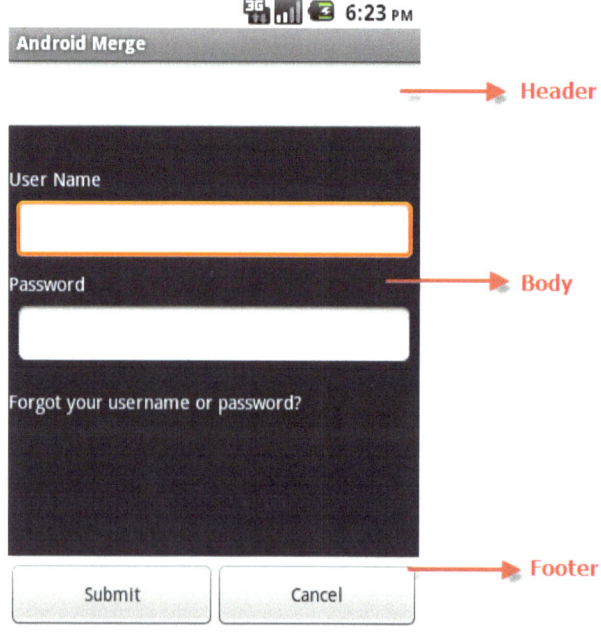

Chapter 9: Shape drawable

This is a generic shape defined in XML.

FILE LOCATION:
res/drawable/*filename*.xml
The filename is used as the resource ID.

RESOURCE REFERENCE:
In XML: @[*package*:]drawable/*filename*

SHAPE ELEMENTS

The shape drawable. This must be the root element.

ATTRIBUTES:
xmlns:android

String. **Required.** Defines the XML namespace, which must be

"http://schemas.android.com/apk/res/android".

android:shape
Keyword. Defines the type of shape. Valid values are:

Value	Description
"rectangle"	A rectangle that fills the containing View. This is the default shape.
"oval"	An oval shape that fits the dimensions of the containing View.
"line"	A horizontal line that spans the width of the containing View. This shape requires the <stroke> element to define the width of the line.
"ring"	A ring shape.

The following attributes are used only when

android:shape="ring":

Attribute	Description
android:innerRadius	*Dimension*. The radius for the inner part of the ring (the hole in the middle), as a dimension value or dimension resource.
android:innerRadiusRatio	*Float*. The radius for the inner part of the ring, expressed as a ratio of the ring's width. For instance, if android:innerRadiusRatio="5", then the inner radius equals the ring's width divided by 5. This value is overridden by android:innerRadius. Default value is 9.
android:thickness	*Dimension*. The thickness of the ring, as a dimension value or dimension resource.
android:thicknessRatio	*Float*. The thickness of the ring, expressed as a ratio of the ring's width. For instance, if android:thicknessRatio ="2", then the thickness equals the ring's width divided by 2. This value is overridden by android:innerRadius. Default value is 3.
android:useLevel	*Boolean*. "true" if this is used as a LevelListDrawable. This should normally be "false" or your shape may not appear.

CORNERS

<corners>

Creates rounded corners for the shape. Applies only when the shape is a rectangle.

ATTRIBUTES	Description
android:radius	*Dimension*. The radius for all corners, as a dimension value. This is overridden for each corner by the following attributes.
android:topLeftRadius	*Dimension*. The radius for the top-left corner, as a dimension value.
android:topRightRadius	*Dimension*. The radius for the top-right corner, as a dimension value.
android:bottomLeftRadius	*Dimension*. The radius for the bottom-left corner, as a dimension value.
android:bottomRightRadius	*Dimension*. The radius for the bottom-right corner, as a dimension value.

> **Note:** Every corner must (initially) be provided a corner radius greater than 1, or else no corners are rounded. If you want specific corners to not be rounded, a work-around is to use android:radius to set a default corner radius greater than 1, but then override each and every corner with the values you really want, providing zero ("0dp") where you don't want rounded corners.

GRADIENT

<gradient>

Specifies a gradient color for the shape

ATTRIBUTES	Description
android:angle	*Integer*. The angle for the gradient, in degrees. 0 is left to right, 90 is bottom to top. It must be a multiple of 45. Default is 0.
android:centerX	*Float*. The relative X-position for the center of the gradient (0 - 1.0).
android:centerY	*Float*. The relative Y-position for the center of the gradient (0 - 1.0).
android:centerColor	*Color*. Optional color that comes between the start and end colors, as a hexadecimal value or color resource.
android:endColor	*Color*. The ending color, as a hexadecimal value or color resource.
android:gradientRadius	*Float*. The radius for the gradient. Only applied when android:type="radial".
android:startColor	*Color*. The starting color, as a hexadecimal value or color resource.
android:type	*Type (linear, radial, sweep)*

android:type

Keyword. The type of gradient pattern to apply. Valid values are:

Value	Description
"linear"	A linear gradient. This is the default.
"radial"	A radial gradient. The start color is the center color.
"sweep"	A sweeping line gradient.

android:useLevel

Boolean. "true" if this is used as a **LevelListDrawable**.

PADDING
<padding>

Padding to apply to the containing View element (this pads the position of the View content, not the shape).

ATTRIBUTES	Description
android:left	*Dimension*. Left padding, as a dimension value or dimension resource.
android:top	*Dimension*. Top padding, as a dimension value or dimension resource.
android:right	*Dimension*. Right padding, as a dimension value or dimension resource.
android:bottom	*Dimension*. Bottom padding, as a dimension value or dimension resource.

SIZE
<size>

The size of the shape

ATTRIBUTES	Description
android:height	*Dimension*. The height of the shape, as a dimension value or dimension resource.
android:width	*Dimension*. The width of the shape, as a dimension value or dimension resource.

Note: The shape scales to the size of the container View proportionate to the dimensions defined here, by default. When you use the shape in an ImageView, you can restrict scaling by setting the android:scaleType to "center".

SOLID

<solid>

A solid color to fill the shape.

ATTRIBUTES	Description
android:color	*Color*. The color to apply to the shape, as a hexadecimal value or color resource.

STROKE

<stroke>

A stroke line for the shape.

Attributes	Description
android:width	*Dimension*. The thickness of the line, as a dimension value or dimension resource.
android:color	*Color*. The color of the line, as a hexadecimal value or color resource.
android:dashGap	*Dimension*. The distance between line dashes, as a dimension value or dimension resource. Only valid if android:dashWidth is set.
android:dashWidth	*Dimension*. The size of each dash line, as a dimension value or dimension resource. Only valid if android:dashGap is set.

Example:

Let's create a gradient box for our textview

1- Create a new project and name it ShapeDrawable
2- Right click on drawable-hdpi on the newly created project to create a **drawable resource**

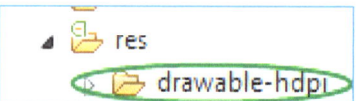

3- Select **New > Others > Android XML File** to open the New Android XML File wizard
4- Under file put **gradient_box.xml**
5- Click **Finish** button

Your new resource folder will look like this:

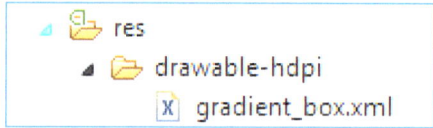

6- Double click on the file **gradient_box.xml** and copy the code bellow

```xml
<?xml version="1.0" encoding="utf-8"?>

<shape xmlns:android="http://schemas.android.com/apk/res/android"

    android:shape="rectangle">

    <gradient

        android:startColor="#FFFF0000"

        android:endColor="#80FF00FF"

        android:angle="45"/>

    <padding android:left="7dp"

        android:top="7dp"

        android:right="7dp"

        android:bottom="7dp" />

    <corners android:radius="8dp" />
```

```
</shape>
```

7- Go into your main.xml file and add the following line of code to your TextView

android:background="@drawable/gradient_box"

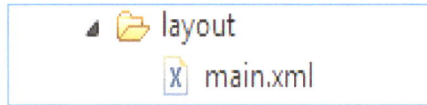

Your textview will look like the following:

```
<TextView
  android:background="@drawable/gradient_box"
    android:layout_width="wrap_content"
    android:layout_height="wrap_content"
    android:text="@string/hello"
    />
```

8- Run the application

The result should look like this :

Create Gradient Button

You can create buttons with ShapeDrawable.

In this example you will create 2 buttons with different colors and you can use it instead of the default android button.

1- Create a new project and name it **Gradient Button**
2- Go in **res/values** and create the style.xml file
3- Copy the following code in style.xml

```xml
<?xml version="1.0" encoding="utf-8"?>
<resources>
<style name="btn_style">
        <item name="android:textColor">#FFFFFF</item>
        <item name="android:gravity">center</item>
        <item name="android:textSize">20sp</item>
        <item name="android:Layout_margin">3dp</item>
        <item name="android:textStyle">bold</item>
        <item name="android:shadowColor">#000000</item>
        <item name="android:shadowDx">0</item>
        <item name="android:shadowDy">-1</item>
        <item name="android:shadowRadius">1</item>
    </style>
</resources>
```

4- Go in **res/layout** directory and open main.xml file

Add the following xml code in main.xml

```xml
<?xml version="1.0" encoding="utf-8"?>
<LinearLayout
xmlns:android="http://schemas.android.com/apk/res/android"
    android:orientation="vertical"
    android:layout_width="fill_parent"
    android:layout_height="fill_parent">
    <Button android:text="Gradient Button1"
    android:layout_marginTop="5sp"
    android:id="@+id/button1"
    android:layout_width="fill_parent"
    android:layout_height="wrap_content"
    android:background="@drawable/btn_gradient1"
    style="@style/btn_style" >
    </Button>

    <Button android:text="Gradient Button2"
    android:layout_marginTop="5sp"
    android:id="@+id/button2"
    android:layout_width="fill_parent"
```

```
        android:layout_height="wrap_content"
        android:background="@drawable/btn_gradient"
        style="@style/btn_style">
    </Button>
</LinearLayout>
```

Note: The attribute android: background = "@drawable/btn_gradient1" allows you to add a drawable resource as a background image.

5- Create a new drawable resource name it **btn_gradient** (button 1) and copy the code bellow

```xml
<?xml version="1.0" encoding="utf-8"?>
<selector xmlns:android="http://schemas.android.com/apk/res/android">
    <item android:state_pressed="true" >
        <shape>
            <solid
                android:color="#2c68e7" />
            <stroke
                android:width="1dp"
                android:color="#2c68e7" />
          <corners
                android:radius="4dp" />
            <padding
                android:left="10dp"
                android:top="10dp"
                android:right="10dp"
                android:bottom="10dp" />
        </shape>
    </item>
    <item>
        <shape>
            <gradient
                android:startColor="#859dbc"
                android:endColor="#4a6c9b"
                android:angle="270" />
            <stroke
                android:width="1dp"
                android:color="#375075" />
            <corners
                android:radius="4dp" />
            <padding
                android:left="10dp"
                android:top="10dp"
                android:right="10dp"
                android:bottom="10dp" />
        </shape>
    </item>
</selector>
```

6- Create a new drawable resource name it **btn_gradient1** (button 2) and copy the code
    bellow

```xml
<?xml version="1.0" encoding="utf-8"?>
<selector xmlns:android="http://schemas.android.com/apk/res/android">
    <item android:state_pressed="true" >
        <shape>
            <solid
                    android:color="#449def" />
            <stroke
                android:width="1dp"
                android:color="#2f6699" />
            <corners
                android:radius="3dp" />
            <padding
                android:left="10dp"
                android:top="10dp"
                android:right="10dp"
                android:bottom="10dp" />
        </shape>
    </item>
    <item>
        <shape>
            <gradient
                android:startColor="#449def"
                android:endColor="#2f6699"
                android:angle="270" />
            <stroke
                android:width="1dp"
                android:color="#2f6699" />
            <corners
                android:radius="4dp" />
            <padding
                android:left="10dp"
                android:top="10dp"
                android:right="10dp"
                android:bottom="10dp" />
        </shape>
    </item>
</selector>
```

The result should look like this :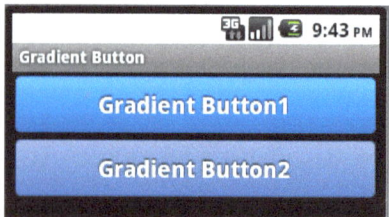

> Note: It's possible to create bars with drawable resources to bring some look and feel to your android applications (eg topbar).

Chapter 10: Android Icon

Creating a unified look and feel throughout a user interface adds value to your product. Streamlining the graphic style will also make the UI seem more professional to users.
This chapter provides information to help you create icons for various parts of your application's user interface that match the general styles used by the Android 2.x framework.

Launcher Icons

A launcher icon is a graphic that represents your application. Launcher icons are used by Launcher applications and appear on the user's Home screen. Launcher icons can also be used to represent shortcuts into your application (for example, a contact shortcut icon that opens detail information for a contact).

Note: You should create separate icons for all generalized screen densities, including low-, medium-, high-, and extra-high-density screens. This ensures that your icons will display properly across the range of devices on which your application can be installed.

Goals of the Launcher Icon

Promote the brand and tell the story of the app.

Help users discover the app in Android Market.

Function well in the Launcher.

Size and Format

Launcher icons should be 32-bit PNGs with an alpha channel for transparency. The finished launcher icon dimensions corresponding to a given generalized screen density are shown in the table below.

	ldpi (120 dpi) (Low density screen)	mdpi (160 dpi) (Medium density screen)	hdpi (240 dpi) (High density screen)	xhdpi (320 dpi) (Extra-high density screen)
Launcher Icon Size	36 x 36 px	48 x 48 px	72 x 72 px	96 x 96 px

You can also include a few pixels of padding in launcher icons to maintain a consistent visual weight with adjacent icons. For example, a 96 x 96 pixel xhdpi launcher icon can contain an 88 x 88 pixel shape with 4 pixels on each side for padding. This padding can also be used to make room for a subtle drop shadow, which can help ensure that launcher icons are legible across on any background color.

Menu Icons

Menu icons are graphical elements placed in the options menu shown to users when they press the Menu button. They are drawn in a flat-front perspective and in greyscale. Elements in a menu icon must not be visualized in 3D or perspective.

> **Note:** Final art must be exported as a transparent PNG file. Do not include a background color.

Menu icons can use a variety of shapes and forms and must be scaled and positioned inside the asset to create consistent visual weight with other icons.

Menu icons are flat, pictured face on, and greyscale.

In order to indicate the recommended size for the icon, each example in includes three different guide rectangles:

The red box is the bounding box for the full asset.

The blue box is the recommended bounding box for the actual icon. The icon box is sized smaller than the full asset box to allow for varying icon shapes while maintaining a consistent visual weight.

The orange box is the recommended bounding box for the actual icon when the content is square. The box for square icons is smaller than that for other icons to establish a consistent visual weight across the two types.

Menu icon dimensions for high-density (hdpi) screens:
1. Full Asset: 72 x 72 px
2. Icon: 48 x 48 px
3. Square Icon: 44 x 44 px

Menu icon dimensions for medium-density (mdpi) screens:
1. Full Asset: 48 x 48 px
2. Icon: 32 x 32 px
3. Square Icon: 30 x 30 px

Menu icon dimensions for low-density (ldpi) screens:
1. Full Asset: 36 x 36 px
2. Icon: 24 x 24 px
3. Square Icon: 22 x 22 px

Action Bar Icons

Action Bar icons are graphical elements placed in the Action Bar representing individual action items.

Size and format

Action Bar icons should be 32-bit PNGs with an alpha channel for transparency. The finished action bar icon dimensions, corresponding to a given generalized screen density, are shown in the table below.

	ldpi (120 dpi) (Low density screen)	mdpi (160 dpi) (Medium density screen)	hdpi (240 dpi) (High density screen)	xhdpi (320 dpi) (Extra-high density screen)
Action Bar Icon Size	18 x 18 px	24 x 24 px	36 x 36 px	48 x 48 px

Status Bar Icons

Status bar icons are used to represent notifications from your application in the status bar.

> **Warning:** The style and dimensions of status bar icons have changed dramatically in Android 3.0 and 2.3 compared to previous versions. **To provide support for all Android versions**, developers should:

Place status bar icons for Android 3.0 and later in the drawable-xhdpi-v11, drawable-hdpi-v11, drawable-mdpi-v11, and drawable-ldpi-v11 directories.

Place status bar icons for Android 2.3 in the drawable-xhdpi-v9, drawable-hdpi-v9, drawable-mdpi-v9, and drawable-ldpi-v9 directories.

Place status bar icons for previous versions in drawable-xhdpi, drawable-hdpi, drawable-mdpi, and drawable-ldpi directories

Size and positioning (Android 2.3)

Status bar icons should use simple shapes and forms and those must be scaled and positioned inside the final asset.

Figure bellow illustrates various ways of positioning the icon inside the asset. You should size the icons *smaller than the actual bounds of the asset*. Status bar icons may vary in width, but only minimally.

In order to indicate the recommended size for the icon, each example includes two different guide rectangles:

The red box is the bounding box for the full asset.
The blue box is the recommended bounding box for the actual icon. The icon box is sized smaller vertically than the full asset box to allow for varying icon shapes while maintaining a consistent visual weight.

Status bar icon dimensions for high-density (hdpi) screens:
1. Full Asset: 24w x 38h px (preferred, width may vary)
2. Icon: 24w x 24h px (preferred, width may vary)

Status bar icon dimensions for medium-density (mdpi) screens:
1. Full Asset: 16w x 25 px (preferred, width may vary)
2. Icon: 16w x 16w px (preferred, width may vary)

Status bar icon dimensions for low-density (ldpi) screens:
1. Full Asset: 12w x 19h px (preferred, width may vary)
2. Icon: 12w x 12h px (preferred, width may vary)

For android 3.0 and later

The design for status bar (notification) icons has been revised in Android 3.0. Status bar icons used in Android 3.0 and later are easier to create, and they allow for more flexible presentation in a variety of situations:

Status bar icons are composed simply of **white pixels on a transparent backdrop**, with alpha blending used for smooth edges and internal texture where appropriate.
Icons are square icon contents should **fill the available space**, although a small amount of internal padding can help maintain balance across status bar icons.

Size and format

Status bar icons should be 32-bit PNGs with an alpha channel for transparency. The finished status bar icon dimensions corresponding to a given generalized screen density are shown in the table below.

> **Note:** The system will shrink and dim status bar icons to minimize distractions, allowing users to focus on the foreground activity.

	ldpi (120 dpi) (Low density screen)	mdpi (160 dpi) (Medium density screen)	hdpi (240 dpi) (High density screen)	xhdpi (320 dpi) (Extra-high density screen)
Status Bar Icon Size (Android 3.0 and Later)	18 x 18 px	24 x 24 px	36 x 36 px	48 x 48 px

Tab Icons

Tab icons are graphical elements used to represent individual tabs in a multi-tab interface. Each tab icon has two states: unselected and selected.

> **Warning:** The style of tab icons has changed drastically in Android 2.0 compared to previous versions. **To provide support for all Android versions**, developers should:
> 1. Place tab icons for Android 2.0 and higher in the drawable-hdpi-v5, drawable-mdpi- v5, and drawable-ldpi-v5 directories.
>
> 2. Place tab icons for previous versions in drawable-hdpi, drawable-mdpi, and drawable-ldpi directories.
>
> 3. Set android:targetSdkVersion to 5 or higher in the **<uses-sdk>** in the application manifest. This will inform the system that it should render tabs using the new tab style

Android 2.0 through Android 2.3

Size and positioning

Tab icons should use simple shapes and forms and those must be scaled and positioned inside the final asset.

Figure bellow illustrates various ways of positioning the icon inside the asset. You should size the icons *smaller than the actual bounds of the asset*.

In order to indicate the recommended size for the icon, each example includes three different guide rectangles:

The red box is the bounding box for the full asset.
The blue box is the recommended bounding box for the actual icon. The icon box is sized smaller than the full asset box to allow for special icon treatments.
The orange box is the recommended bounding box for the actual icon when the content is square. The box for square icons is smaller than that for other icons to establish a consistent visual weight across the two types.

Tab icon dimensions for high-density (hdpi) screens:
1. Full Asset: 48 x 48 px
2. Icon: 42 x 42 px

Tab icon dimensions for medium-density (mdpi) screens:
1. Full Asset: 32 x 32 px
2. Icon: 28 x 28 px

Tab icon dimensions for low-density (ldpi) screens:
1. Full Asset: 24 x 24 px
2. Icon: 22 x 22 px

Dialog Icons

Dialog icons are shown in pop-up dialog boxes that prompt the user for interaction. They use a light gradient and inner shadow in order to stand out against a dark background.

Low density screen (ldpi)	Medium density screen (mdpi)	High density screen (hdpi)
24 x 24 px	32 x 32 px	48 x 48 px

Part III

Creating Custom Control

Chapter 11: Custom Control

Introducing NinePatch Drawable

A NinePatchDrawable graphic is a stretchable bitmap image, which Android will automatically resize to accommodate the contents of the View in which you have placed it as the background. An example use of a NinePatch is the backgrounds used by standard Android buttons, buttons must stretch to accommodate strings of various lengths. A NinePatch drawable is a standard PNG image that includes an extra 1-pixel-wide border. It must be saved with the extension *.9.png*, and saved into the **res/drawable/** directory of your project.

The border is used to define the stretchable and static areas of the image. You indicate a stretchable section by drawing one (or more) 1-pixel-wide black line(s) in the left and top part of the border (the other border pixels should be fully transparent or white). You can have as many stretchable sections as you want: their relative size stays the same, so the largest sections always remain the largest.

You can also define an optional drawable section of the image (effectively, the padding lines) by drawing a line on the right and bottom lines. If a View object sets the NinePatch as its background and then specifies the View's text, it will stretch itself so that all the text fits inside only the area designated by the right and bottom lines (if included). If the padding lines are not included, Android uses the left and top lines to define this drawable area.

To clarify the difference between the different lines, the left and top lines define which pixels of the image are allowed to be replicated in order to stretch the image. The bottom and right lines define the relative area within the image that the contents of the View are allowed to lie within.

Here is a sample NinePatch file used to define a button:

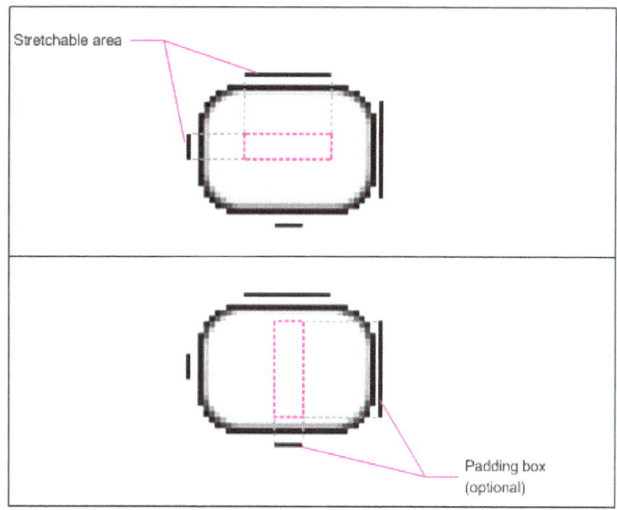

This NinePatch defines one stretchable area with the left and top lines and the drawable area with the bottom and right lines. In the top image, the dotted grey lines identify the regions of the image that will be replicated in order to stretch the image. The pink rectangle in the bottom image identifies the region in which the contents of the View are allowed.

If the contents don't fit in this region, then the image will be stretched

The Draw 9-patch tool offers an extremely handy way to create your NinePatch images, using a WYSIWYG graphics editor. It even raises warnings if the region you've defined for the stretchable area is at risk of producing drawing artifacts as a result of the pixel replication.

Custom Styled button

```
<Button android:text="Android"
android:id="@+id/button1"
style="@style/ButtonText"
android:background="@drawable/a9p_09_11_00610_9"
android:layout_width="fill_parent">
</Button>
```

You can see that the button is not resized properly and android automatically stretch the button but it doesn't look like you would like it to be.

You can correct this using android NinePatch

1- Go to **File** > **Open 9-patch...** and locate the button image to correct

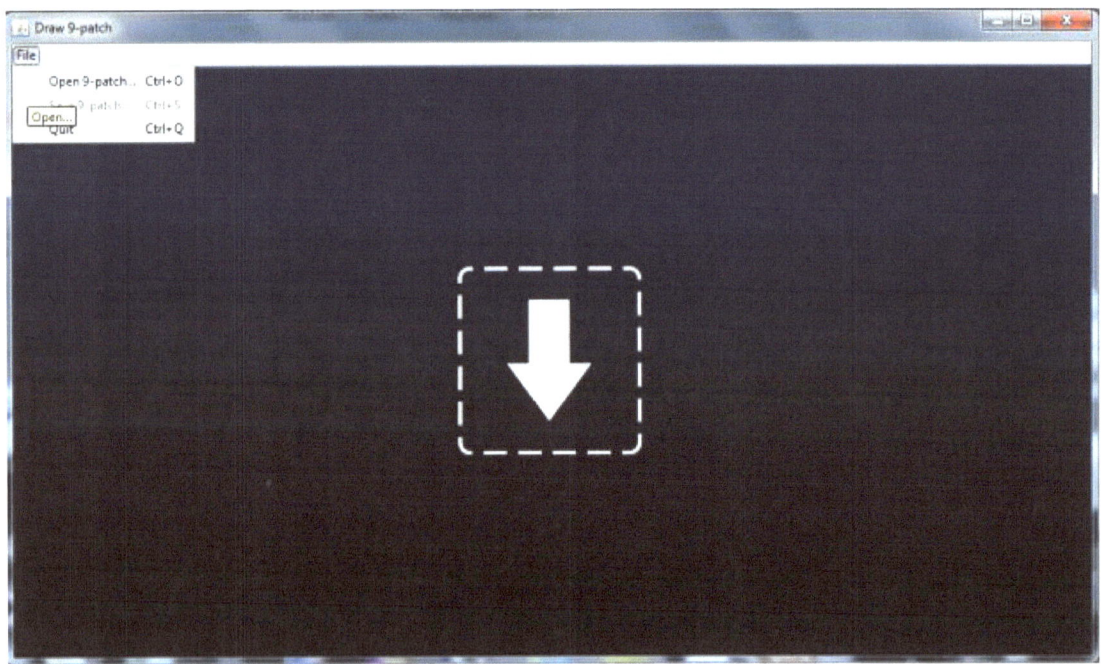

2- A new window appears with the button image

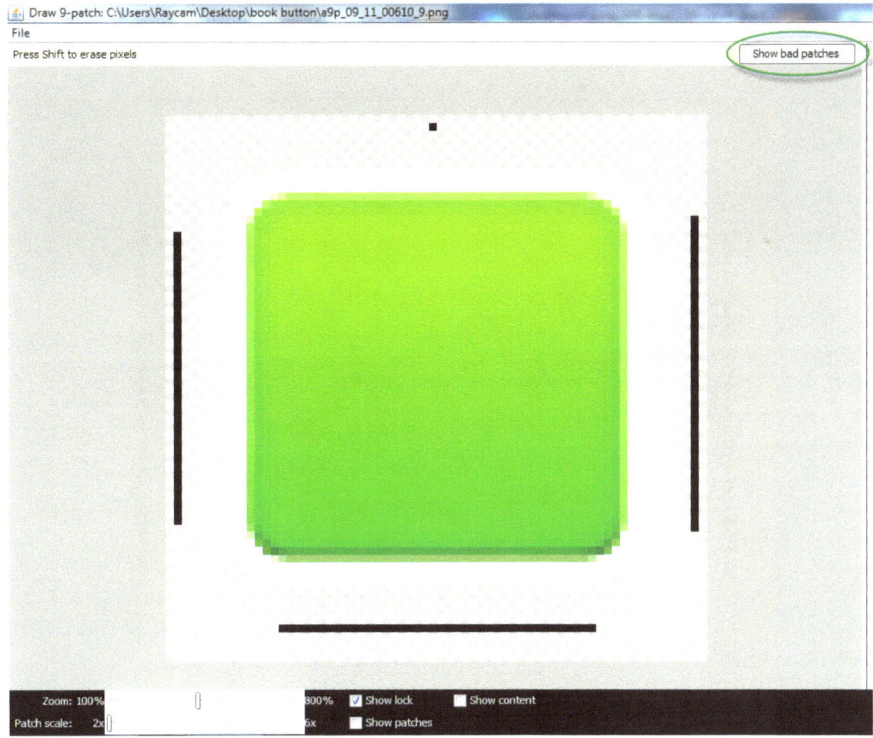

3- Click **on show bad patches** located at the upper right corner

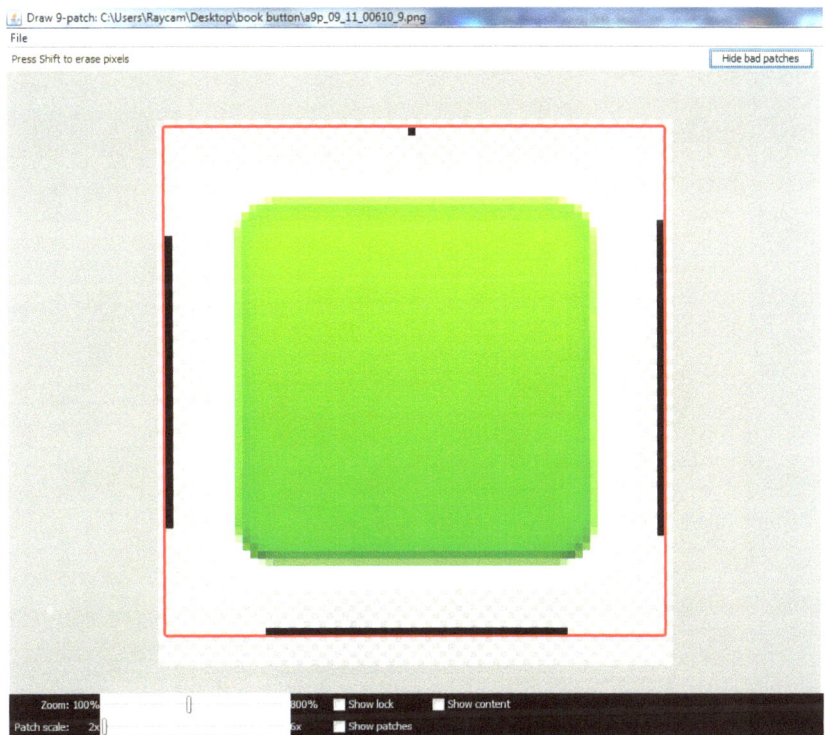

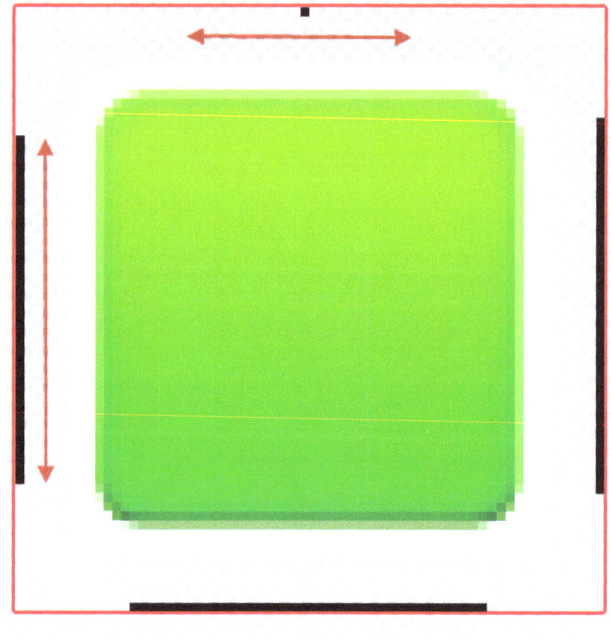

 : Represent the scalable area

4- Left click in red rounded region to draw pixel, you can see in the right pane a preview of your button corrected.

Congratulations there is no batch patch!

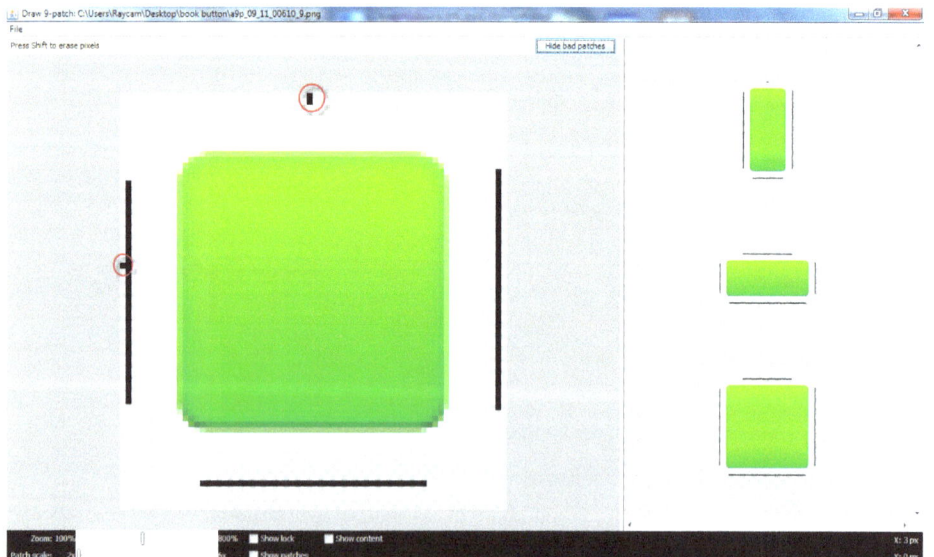

5- Go to **File > Save 9-patch...**

Your new button image is changed to ninepatch

Android automatically add the **.9.png** extension

6- Now copy and paste your custom button image in **res/drawable** folder

In the button widget xml add the name of the new saved button **"button9patched"** without the **.9.png** extension like in this code

```
<Button android:text="Glossy"
android:id="@+id/button1"
style="@style/ButtonText"
android:background="@drawable/button9patched"
android:layout_width="fill_parent"></Button>
```

7- Run the application

Button example

Here some 9-patches images, you can draw your own and resize them using **9patch drawable** software located in android_sdk_dir/tools/draw9patch.
Or download free ones at http://android9patch.blogspot.com/

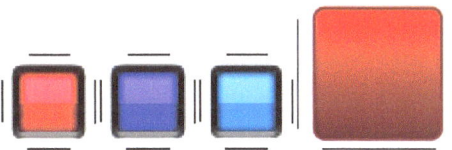

1- Create a new project and name it custom Button

2- Go in **res/values** and create the style.xml file and copy the following code in your new created file

```
<?xml version="1.0" encoding="utf-8"?>
<resources>
```

```xml
<string name="hello"></string>
<string name="app_name">Custom Button</string>
<style name="ButtonText">
<item name="android:layout_width">wrap_content</item>
<item name="android:layout_height">wrap_content</item>
<item name="android:textColor">#ffffff</item>
<item name="android:gravity">center</item>
<item name="android:layout_margin">3dp</item>
<item name="android:textSize">30dp</item>
<item name="android:shadowColor">#000000</item>
<item name="android:shadowDx">1</item>
<item name="android:shadowDy">1</item>
<item name="android:shadowRadius">2</item>
<item name="android:paddingLeft">15dp</item>
<item name="android:paddingRight">15dp</item>
</style>

</resources>
```

3- Create a vertical LinearLayout resource

This time add custom buttons with the attribute **android:background = "@drawable/a9p_09_11_00108"** and apply a style to each button.

```xml
<LinearLayout
xmlns:android="http://schemas.android.com/apk/res/android"
android:layout_width="fill_parent"
android:layout_height="fill_parent"
android:orientation="vertical">

</LinearLayout>
```

4- Add some buttons controls

```xml
<Button android:text="Glossy"
  android:id="@+id/button1"
  style="@style/ButtonText"
  android:background="@drawable/a9p_09_11_00103"
  android:layout_width="fill_parent"></Button>

<Button android:text="Glossy"
  android:id="@+id/button3"
  style="@style/ButtonText"
  android:background="@drawable/a9p_09_11_00108"
  android:layout_width="fill_parent"></Button>

<Button android:text="Glossy"
  android:id="@+id/button4"
  style="@style/ButtonText"
```

```
        android:background="@drawable/a9p_09_11_00102"
        android:layout_width="fill_parent"></Button>

    <Button android:text="Matte"
        android:id="@+id/button2"
        style="@style/ButtonText"
        android:background="@drawable/a9p_09_11_00080"
        android:layout_width="fill_parent"></Button>
```

Result should look something like this:

Create custom font

Android provide default font, but if you want to add a touch a little different to your application by adding pretty font, here are the steps to follow:

1- Create a new project name it CustomFont

2- Create the **/fonts** folder in the **/assets** directory.

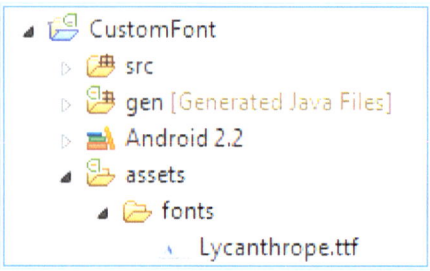

3- Copy the Lycanthrophe.ttf (downloaded from http://www.sinisterfonts.com)in the **/fonts** directory

4- Copy the following code in the main.xml

```xml
<?xml version="1.0" encoding="utf-8"?>
<LinearLayout
xmlns:android="http://schemas.android.com/apk/res/android"
    android:orientation="vertical"
    android:layout_width="fill_parent"
    android:layout_height="fill_parent"
    >
    <TextView
      android:paddingTop="20sp"
      android:id="@+id/txt"
      android:layout_width="fill_parent"
      android:layout_height="wrap_content"
      android:text="Horror font"
      android:textSize="30sp"
    />
</LinearLayout>
```

5- Use the following Java code to bind the font with the UI widget wanting to display the custom typeface

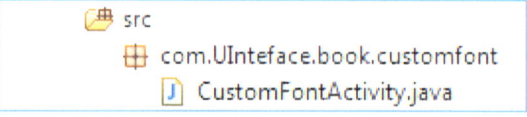

```
package com.UInteface.book.customfont;

import android.app.Activity;
import android.os.Bundle;
import android.graphics.Typeface;
import android.widget.TextView;


public class CustomFontActivity extends Activity {
    /** Called when the activity is first created. */
    @Override
    public void onCreate(Bundle savedInstanceState) {
        super.onCreate(savedInstanceState);
        setContentView(R.layout.main);

        TextView txt=(TextView)findViewById(R.id.txt);
        Typeface myNewFace=Typeface.createFromAsset(getAssets(),
"fonts/Lycanthrope.ttf" );
        txt.setTypeface(myNewFace);

    }
}
```

6- Run the application

The final result should look like this:

Create Custom Title Bar

1- Create a new project and name it Custom Title Bar

2- Go in **res/values** and create the style.xml file and copy the following code in style.xml

```xml
<?xml version="1.0" encoding="utf-8"?>
<resources>
    <style name="text_title">
        <item name="android:textColor">#FFF</item>
        <item name="android:textStyle">bold</item>
        <item name="android:textSize">24sp</item>
        <item name="android:gravity">center</item>
        <item name="android:paddingLeft">5sp</item>
        <item name="android:shadowDx">1.0</item>
        <item name="android:shadowDy">1.0</item>
        <item name="android:shadowRadius">1</item>
        <item name="android:shadowColor">#2c68e7</item>
        <item name="android:background">#449def</item>
    </style>
</resources>
```

3- Go in **res/values** and create attrs.xml , copy the following code

```xml
<?xml version="1.0" encoding="utf-8"?>
<resources>
  <attr name="textTitle" format="reference" />
</resources>
```

4- Go in **res/layout** directory open the main.xml file and add the following xml code in it

```xml
<?xml version="1.0" encoding="utf-8"?>
<LinearLayout
xmlns:android="http://schemas.android.com/apk/res/android"
    android:orientation="vertical"
    android:layout_width="fill_parent"
    android:layout_height="fill_parent">

        <TextView
        android:id="@+id/customTitle"
        android:layout_width="fill_parent"
        android:layout_height="wrap_content"
        style="?textTitle"
        android:text="My Title Bar"
         />
</LinearLayout>
```

5- Go in **res/values** and create a new file called **MyTheme.xml** and copy the following code

```xml
<?xml version="1.0" encoding="utf-8"?>
<resources>
    <style name="Theme" parent="android:Theme">
        <item name="android:windowTitleSize">20dip</item>
    </style>

     <style name="MyTheme">
        <item name="textTitle">@style/text_title</item>
        <item name="android:windowNoTitle">false</item>
     </style>
</resources>
```

To change the theme in the manifest see chapter 5 "Applying styles and themes to the UI"

Result should look like this:

Oops it's a little confuse having two titles!

To hide the Android Title bar for your activities, go in Theme.xml file and add this line of code

```xml
<item name="android:windowNoTitle">true</item>
```

Result should look something like this:

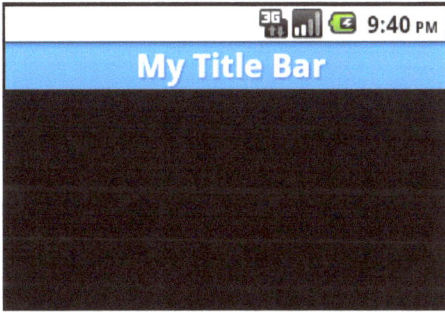

Case-Study Game Interface

The final output will be like below prototype

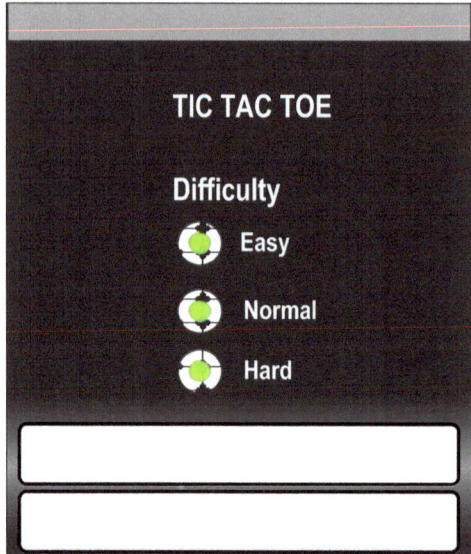

Difficulty level

1- Create a new project and name it gamelevel

2- Add your Images in **res/drawable** folder

3- Create a vertical LinearLayout resource, add a background image

```xml
<?xml version="1.0" encoding="utf-8"?>
<LinearLayout
xmlns:android="http://schemas.android.com/apk/res/android"
    android:orientation="vertical"
    android:layout_width="fill_parent"
    android:layout_height="fill_parent"
    android:background="@drawable/bitmapdraw"
  >
</LinearLayout>
```

4- Between <LinearLayout> and </LinearLayout>, add a background, an ImageView, a TextView.

```xml
<ImageView
android:src="@drawable/tictac1"
android:layout_width="200px"
android:layout_height="100px"
android:layout_gravity="center"/>
<TextView android:text="Difficulty"
```

```
android:id="@+id/TextView01"
android:layout_width="fill_parent"
android:layout_height="wrap_content"
android:gravity="center"
android:paddingTop="20sp"
android:textSize="20sp"></TextView>
```

5- Then you will include a RelativeLayout that host the radiobutton, remember that a radiobutton must be in a radiogroup.

To position the radiogroup in the center, set **android:paddingLeft="100sp"**

```
<RelativeLayout
android:orientation="horizontal"
android:layout_width="fill_parent"
android:layout_height="wrap_content"
>

<RadioGroup
        android:orientation="vertical"
        android:layout_width="fill_parent"
        android:layout_height="wrap_content"
        android:paddingLeft="100sp"
        >
        <RadioButton
                android:layout_width="wrap_content"
                android:layout_height="wrap_content"
                android:textColor="#FD0303"
                android:text="Easy" />

        <RadioButton
                android:layout_width="wrap_content"
                android:layout_height="wrap_content"
            android:text="Normal" />
        <RadioButton
        android:textColor="#4CFF00"
        android:text="Hard" />
</RadioGroup>
</RelativeLayout>
```

6- Add a vertical LinearLayout that will contain the buttons **Start** et **Scoreboard**

```
<LinearLayout
android:orientation="vertical"
android:layout_width="fill_parent"
android:layout_height="wrap_content"
android:gravity="center"
android:layout_marginTop="12sp"
>
```

```
<Button android:text="Start" android:id="@+id/Button01"
android:layout_width="fill_parent" android:layout_height="wrap_content"
android:layout_marginTop="12sp"></Button>

<Button android:text="Scoreboard" android:id="@+id/Button01"
android:layout_width="fill_parent" android:layout_height="wrap_content"
>
</Button>
</LinearLayout>
```

> **Note:** **android:layout_marginTop="12sp** *leaves* a space between the buttons and radiobutton

Result should look something like this:

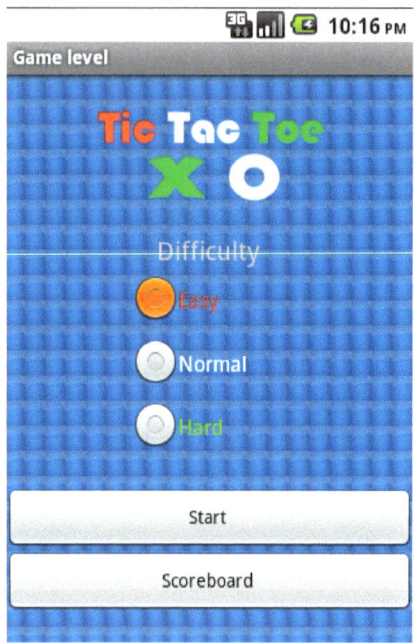

If you wish to see your screen display only in portrait mode, you must add an entry in *AndroidManifest.xml* saying that app runs only in portrait mode.

```
<?xml version="1.0" encoding="utf-8"?>
<manifest xmlns:android="http://schemas.android.com/apk/res/android"
      package="com.UInterface.book.gamelevel"
      android:versionCode="1"
      android:versionName="1.0">
    <uses-sdk android:minSdkVersion="8" />

      <application android:icon="@drawable/icon"
      android:label="@string/app_name">
        <activity android:name=".GamelevelActivity"
                android:label="@string/app_name"
```

```
        android:screenOrientation="portrait">  ⬅
        <intent-filter>
            <action android:name="android.intent.action.MAIN" />
            <category android:name="android.intent.category.LAUNCHER" />
        </intent-filter>
    </activity>

  </application>
</manifest>
```

Game opening screen

The final output will be like below prototype

1- Create a new project and name it TicTacToe

2- Add your Images in **res/drawable** folder

3- Go in **res/values** and create the **style.xml** file and copy the following code in your new created file

```xml
<?xml version="1.0" encoding="utf-8"?>
<resources>
<string name="hello">Hello World, GamescreenActivity!</string>
<string name="app_name">Game screen</string>

<style name="ButtonText">
<item name="android:layout_width">wrap_content</item>
<item name="android:layout_height">wrap_content</item>
<item name="android:textColor">#ffffff</item>
<item name="android:gravity">center</item>
<item name="android:layout_margin">3dp</item>
<item name="android:textSize">30dp</item>
<item name="android:shadowColor">#000000</item>
<item name="android:shadowDx">1</item>
<item name="android:shadowDy">1</item>
<item name="android:shadowRadius">2</item>
<item name="android:paddingLeft">15dp</item>
<item name="android:paddingRight">15dp</item>
</style>
</resources>
```

4- Go in **res/layout** directory open the main.xml file.
   Create a LinearLayout between **<LinearLayout>** and **</LinearLayout>**, add the properties android layout_width and android:layout_height to fill_parent then add a background image, an ImageView and customized buttons.

```xml
<?xml version="1.0" encoding="utf-8"?>
<LinearLayout
  xmlns:android="http://schemas.android.com/apk/res/android"
  android:layout_width="fill_parent"
  android:layout_height="fill_parent"
  android:orientation="vertical"
  android:background="@drawable/bitmapdraw"
  >
<ImageView
 android:src="@drawable/imgview"
 android:layout_width="180px"
 android:layout_height="180px"
 android:layout_gravity="center" />

<Button android:text="Start Game"
 android:id="@+id/button1"
 style="@style/ButtonText"
 android:background="@drawable/a9p_09_11_00103"
 android:layout_gravity="center"
 android:layout_width="250px">
</Button>

<Button android:text="Options"
 android:id="@+id/button3"
 style="@style/ButtonText"
 android:background="@drawable/a9p_09_11_00103"
 android:layout_gravity="center"
 android:layout_width="250px">
</Button>

<Button android:text="Help"
 android:id="@+id/button4"
 style="@style/ButtonText"
 android:background="@drawable/a9p_09_11_00103"
 android:layout_gravity="center"
 android:layout_width="250px">
</Button>
</LinearLayout>
```

Note: The attribute **android:layout_gravity="center"** allows you to center the ImageView and the buttons.

Result should look something like this:

You will change your game screen orientation that is to say from portrait to landscape, you only see the first two buttons,the others run off the bottom of the screen.

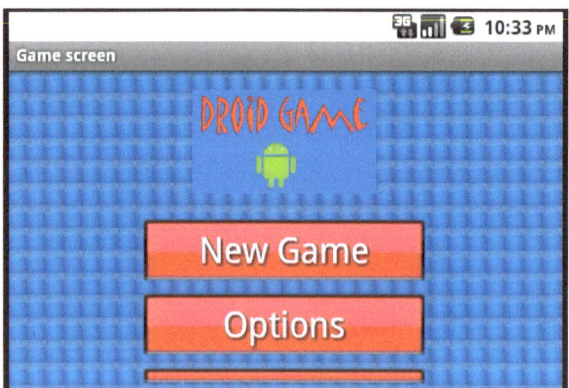

Here is how to correct this

The final output will be like below prototype

1- Create a new **res/layout-land** folder for landscape mode

2- Define a layout and name it ***main.xml***

3- Create a LinearLayout between **<LinearLayout>** and **</LinearLayout>**, add the properties android layout_width and android:layout_height to fill_parent then add a background image, an ImageView and customized buttons.

4- Add your Images in **res/drawable** folder and copy the following code

```
<?xml version="1.0" encoding="utf-8"?>
<LinearLayout
xmlns:android="http://schemas.android.com/apk/res/android"
android:layout_width="fill_parent"
android:layout_height="fill_parent"
android:orientation="vertical"
android:background="@drawable/bitmapdraw"
>
<ImageView
android:src="@drawable/imgview"
android:layout_width="180px"
android:layout_height="180px"
android:layout_gravity="center" />
```

5- You will put the buttons in TableRow

```
<tablelayout
<tableRow>
<Button android:text="Start Game"
android:id="@+id/button1"
style="@style/ButtonText"
android:background="@drawable/a9p_09_11_00103"
android:layout_gravity="center"
```

```
    android:layout_width="250px"></Button>

    <Button android:text="Options"
    android:id="@+id/button3"
    style="@style/ButtonText"
    android:background="@drawable/a9p_09_11_00103"
    android:layout_gravity="center"
    android:layout_width="250px"></Button>
</tableRow>

<tableRow>
<Button android:text="Options"
 android:id="@+id/button3"
 style="@style/ButtonText"
 android:background="@drawable/a9p_09_11_00103"
 android:layout_gravity="center"
 android:layout_width="250px"></Button>

<Button android:text="Help"
android:id="@+id/button4"
style="@style/ButtonText"
android:background="@drawable/a9p_09_11_00103"
android:layout_gravity="center"
android:layout_width="250px"></Button>

</tableRow>
```

6- Run the application

7- Hit **Control-F11** on emulator to change screen orientation to landscape

    Result should look something like this:

Let's localize the application to render it differently in different locales.

To create these resource files

1- Right click in localization select **NEW > OTHERS>Android XML File** to open the New Android XML File wizard

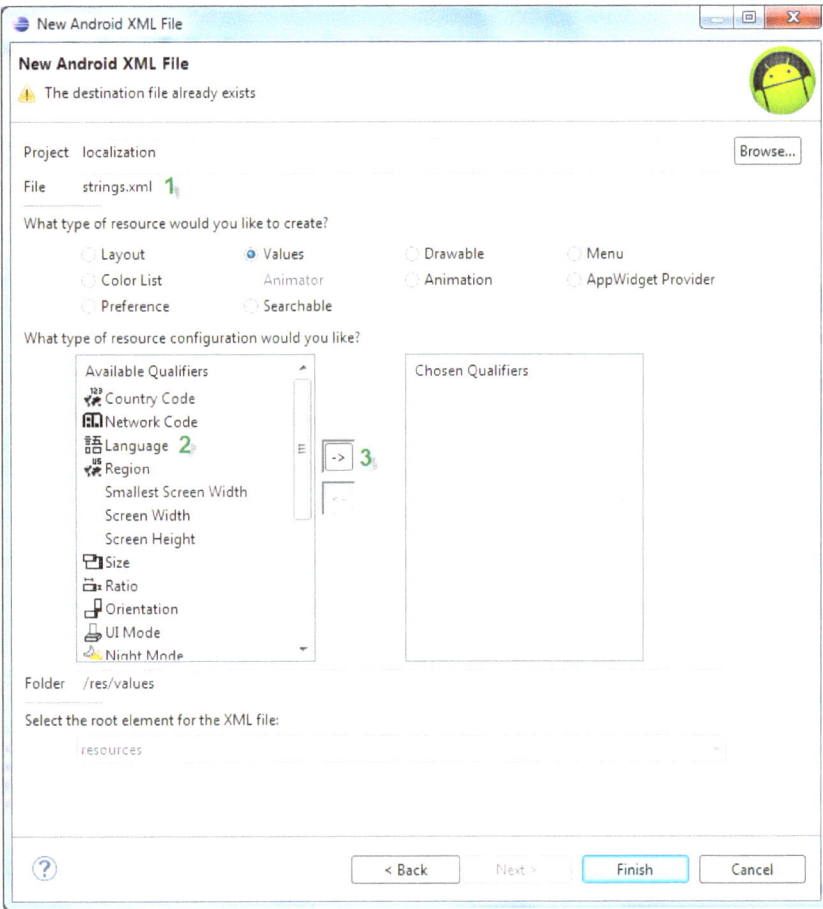

2- Specify the file name **1** and the type of resource **2**

3- Type **fr** in the Language box and click **Finish**

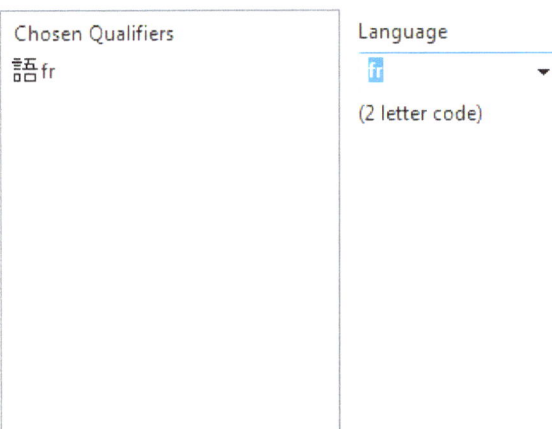

4- A new file, **res/values-fr/strings.xml**, now appears among the project files.

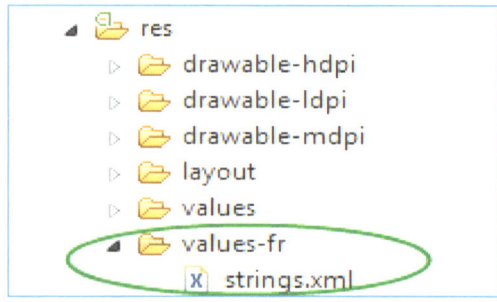

Repeat the steps for es (Spanish) language

You can test in Eclipse to see how your applications will be in different language

   1- Open the main.xml file

   2- Click the eclipse graphical layout

   3- Click the combobox in the upper right and select the desired language see
      screenshot below.

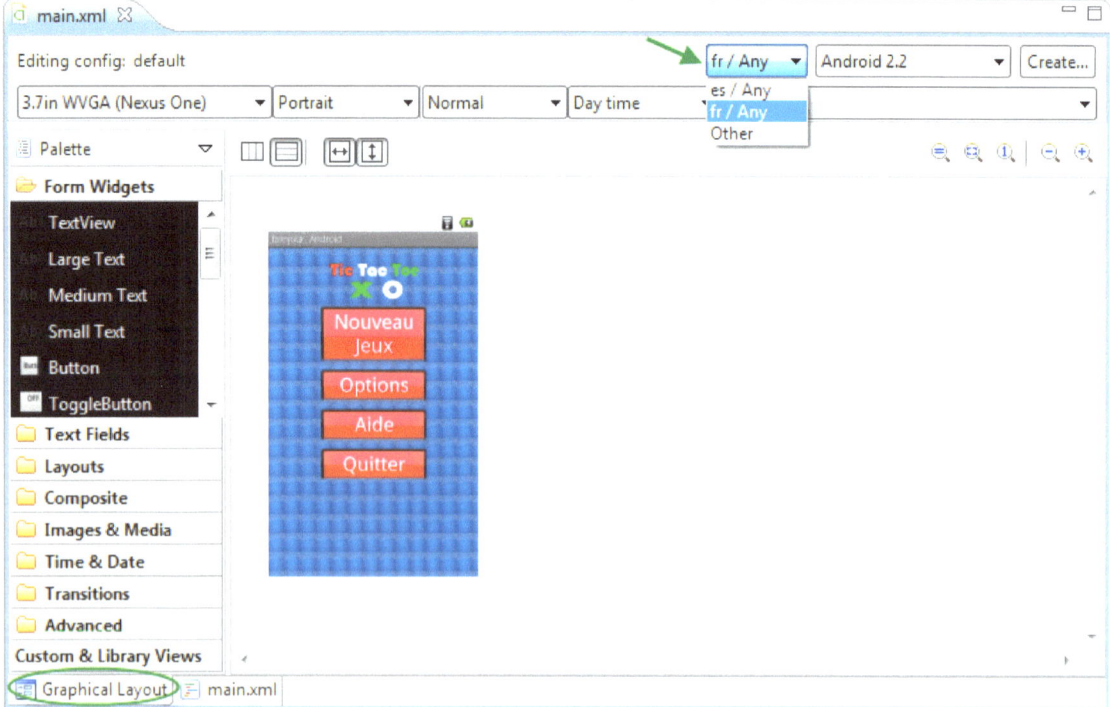

Appendix A: Resource directories

Table 1 Resource directories supported inside project **res/** directory.

Directory	Resource Type
animator/	XML files that define property animations.
anim/	XML files that define tween animations. (Property animations can also be saved in this directory, but the animator/ directory is preferred for property animations to distinguish between the two types.)
color/	XML files that define a state list of colors.
drawable/	Bitmap files (.png, .9.png, .jpg, .gif) or XML files that are compiled into the following drawable resource subtypes: Bitmap files Nine-Patches (re-sizable bitmaps) State lists Shapes Animation drawables Other drawables
layout/	XML files that define a user interface layout.
menu/	XML files that define application menus, such as an Options Menu, Context Menu, or Sub Menu.
raw/	Arbitrary files to save in their raw form. To open these resources with a raw InputStream, call Resources.openRawResource() with the resource ID, which is R.raw.*filename*. However, if you need access to original file names and file hierarchy, you might consider saving some resources in the assets/ directory (instead of res/raw/). Files in assets/ are not given a resource ID, so you can read them only using AssetManager.
values/	XML files that contain simple values, such as strings, integers, and colors. Whereas XML resource files in other res/ subdirectories define a single resource based on the XML filename, files in the values/ directory describe multiple resources. For a file in this directory, each child of the <resources> element defines a single resource. For example, a <string> element creates anR.string resource and a <color> element creates an R.color resource. Because each resource is defined with its own XML element, you can name the file whatever you want and place different resource types in one file. However, for clarity, you might want to place unique resource types in different files. For example, here are some filename

	conventions for resources you can create in this directory: arrays.xml for resource arrays (typed arrays). colors.xml for color values dimens.xml for dimension values. strings.xml for string values. styles.xml for styles..
xml/	Arbitrary XML files that can be read at runtime by calling <u>Resources.getXML()</u>. Various XML configuration files must be saved here, such as a <u>searchable configuration</u>.

Caution: Never save resource files directly inside the res/ directory it will cause a compiler error.

Table 2. Configuration qualifier names.

Configuration	Qualifier Values	Description
MCC and MNC	Examples: mcc310 mcc310-mnc004 mcc208-mnc00 etc.	The mobile country code (MCC), optionally followed by mobile network code (MNC) from the SIM card in the device. For example, mcc310 is U.S. on any carrier, mcc310-mnc004 is U.S. on Verizon, and mcc208-mnc00 is France on Orange. If the device uses a radio connection (GSM phone), the MCC comes from the SIM, and the MNC comes from the network to which the device is connected.
Language and region	Examples: en fr en-rUS fr-rFR fr-rCA etc.	The language is defined by a two-letter ISO 639-1 language code, optionally followed by a two letter ISO 3166-1-alpha-2 region code (preceded by lowercase "r"). The codes are *not* case-sensitive; the r prefix is used to distinguish the region portion. You cannot specify a region alone. This can change during the life of your application if the user changes his or her language in the system settings.
smallestWidth	sw<N>dp Examples: sw320dp sw600dp sw720dp etc.	The fundamental size of a screen, as indicated by the shortest dimension of the available screen area. Specifically, the device's smallestWidth is the shortest of the screen's available height and width (you may also think of it as the "smallest possible width" for the screen). You can use this qualifier to ensure that, regardless of the screen's current orientation, your application's has at least <N> dps of width available for it UI. For example, if your layout requires that its smallest dimension of screen area be at least 600 dp at all times, then you can use this qualifer to create the layout resources, res/layout-sw600dp/. The system will use these resources only when the smallest dimension of

| | | available screen is at least 600dp, regardless of whether the 600dp side is the user-perceived height or width. The smallestWidth is a fixed screen size characteristic of the device; **the device's smallest Width does not change when the screen's orientation changes**.

The smallestWidth of a device takes into account screen decorations and system UI. For example, if the device has some persistent UI elements on the screen that account for space along the axis of the smallestWidth, the system declares the smallestWidth to be smaller than the actual screen size, because those are screen pixels not available for your UI. Thus, the value you use should be the actual smallest dimension *required by your layout* (usually, this value is the "smallest width" that your layout supports, regardless of the screen's current orientation).

Some values you might use here for common screen sizes:

320, for devices with screen configurations such as:

240x320 ldpi (QVGA handset)

320x480 mdpi (handset)

480x800 hdpi (high density handset)

480, for screens such as 480x800 mdpi (tablet/handset).

600, for screens such as 600x1024 mdpi (7" tablet).

720, for screens such as 720x1280 mdpi (10" tablet).

When your application provides multiple resource directories with different values for the smallestWidth qualifier, the system uses the one closest to (without exceeding) the device's smallestWidth.

Added in API level 13. |
| Available width | w<N>dp

Examples:
w720dp
w1024dp
etc. | Specifies a minimum available screen width, in dp units at which the resource should be used—defined by the <N> value. This configuration value will change when the orientation changes between landscape and portrait to match the current actual width.

When your application provides multiple resource directories with different values for this configuration, the system uses the one closest to (without exceeding) the device's current screen width. The value here takes into account screen decorations, so if the device has some persistent UI elements on the left or right edge of the display, it uses a value for the width that is smaller than the real screen size, accounting for these UI elements and reducing the application's available space.

Added in API level 13. |
| Available height | h<N>dp

Examples:
h720dp | Specifies a minimum available screen height, in "dp" units at which the resource should be used—defined by the <N> value. This configuration value will change when the orientation changes |

	h1024dp etc.	between landscape and portrait to match the current actual height. When your application provides multiple resource directories with different values for this configuration, the system uses the one closest to (without exceeding) the device's current screen height. The value here takes into account screen decorations, so if the device has some persistent UI elements on the top or bottom edge of the display, it uses a value for the height that is smaller than the real screen size, accounting for these UI elements and reducing the application's available space. Screen decorations that are not fixed (such as a phone status bar that can be hidden when full screen) are *not* accounted for here, nor are window decorations like the title bar or action bar, so applications must be prepared to deal with a somewhat smaller space than they specify. *Added in API level 13.*
Screen size	small normal large xlarge	• small: Screens that are of similar size to a low-density QVGA screen. The minimum layout size for a small screen is approximately 320x426 dp units. Examples are QVGA low density and VGA high density. • normal: Screens that are of similar size to a medium-density HVGA screen. The minimum layout size for a normal screen is approximately 320x470 dp units. Examples of such screens a WQVGA low density, HVGA medium density, WVGA high density. • large: Screens that are of similar size to a medium-density VGA screen. The minimum layout size for a large screen is approximately 480x640 dp units. Examples are VGA and WVGA medium density screens. • xlarge: Screens that are considerably larger than the traditional medium-density HVGA screen. The minimum layout size for an xlarge screen is approximately 720x960 dp units. In most cases, devices with extra large screens would be too large to carry in a pocket and would most likely be tablet-style devices. *Added in API level 9.* **Note:** Using a size qualifier does not imply that the resources are *only* for screens of that size. If you do not provide alternative resources with qualifiers that better match the current device configuration, the system may use whichever resources are the best match. **Caution:** If all your resources use a size qualifier that is *larger* than the current screen, the system will **not** use them and your application will crash at runtime (for example, if all layout

		resources are tagged with the xlarge qualifier, but the device is a normal-size screen). *Added in API level 4.*
Screen aspect	long notlong	• long: Long screens, such as WQVGA, WVGA, FWVGA • notlong: Not long screens, such as QVGA, HVGA, and VGA *Added in API level 4.* This is based purely on the aspect ratio of the screen (a "long" screen is wider). This is not related to the screen orientation.
Screen orientation	port land	• port: Device is in portrait orientation (vertical) • land: Device is in landscape orientation (horizontal) This can change during the life of your application if the user rotates the screen
Dock mode	car desk	• car: Device is in a car dock • desk: Device is in a desk dock *Added in API level 8.* This can change during the life of your application if the user places the device in a dock. You can enable or disable this mode using UiModeManager.
Night mode	night notnight	• night: Night time • notnight: Day time *Added in API level 8.* This can change during the life of your application if night mode is left in auto mode (default), in which case the mode changes based on the time of day. You can enable or disable this mode using UiModeManager.
Screen pixel density (dpi)	ldpi mdpi hdpi xhdpi nodpi tvdpi	• ldpi: Low-density screens; approximately 120dpi. • mdpi: Medium-density (on traditional HVGA) screens; approximately 160dpi. • hdpi: High-density screens; approximately 240dpi. • xhdpi: Extra high-density screens; approximately 320dpi. *Added in API Level 8* • nodpi: This can be used for bitmap resources that you do not want to be scaled to match the device density. • tvdpi: Screens somewhere between mdpi and hdpi; approximately 213dpi. This is not considered a "primary" density group. It is mostly intended for televisions and most apps shouldn't need it—providing mdpi and hdpi resources is sufficient for most apps and the system will scale them as appropriate. This qualifier was

		introduced with API level 13. There is a 3:4:6:8 scaling ratio between the four primary densities (ignoring the tvdpi density). So, a 9x9 bitmap in ldpi is 12x12 in mdpi, 18x18 in hdpi and 24x24 in xhdpi. If you decide that your image resources don't look good enough on a television or other certain devices and want to try tvdpi resources, the scaling factor is 1.33*mdpi. For example, a 100px x 100px image for mdpi screens should be 133px x 133px for tvdpi. **Note:** Using a density qualifier does not imply that the resources are *only* for screens of that density. If you do not provide alternative resources with qualifiers that better match the current device configuration, the system may use whichever resources are the <u>best match</u>
Touchscreen type	notouch stylus finger	• notouch: Device does not have a touchscreen. • stylus: Device has a resistive touchscreen that's suited for use with a stylus. • finger: Device has a touchscreen.
Keyboard availability	keysexposed keyshidden keyssoft	• keysexposed: Device has a keyboard available. If the device has a software keyboard enabled (which is likely), this may be used even when the hardware keyboard is*not* exposed to the user, even if the device has no hardware keyboard. If no software keyboard is provided or it's disabled, then this is only used when a hardware keyboard is exposed. • keyshidden: Device has a hardware keyboard available but it is hidden *and* the device does *not* have a software keyboard enabled. • keyssoft: Device has a software keyboard enabled, whether it's visible or not. If you provide keysexposed resources, but not keyssoft resources, the system uses the keysexposed resources regardless of whether a keyboard is visible, as long as the system has a software keyboard enabled. This can change during the life of your application if the user opens a hardware keyboard.
Primary text input method	nokeys qwerty 12key	• nokeys: Device has no hardware keys for text input. • qwerty: Device has a hardware qwerty keyboard, whether it's visible to the user or not.

		• 12key: Device has a hardware 12-key keyboard, whether it's visible to the user or not.
Navigation key availability	navexposed navhidden	• navexposed: Navigation keys are available to the user. • navhidden: Navigation keys are not available (such as behind a closed lid). This can change during the life of your application if the user reveals the navigation keys.
Primary non-touch navigation method	nonav dpad trackball wheel	• nonav: Device has no navigation facility other than using the touch screen. • dpad: Device has a directional-pad (d-pad) for navigation. • trackball: Device has a trackball for navigation. • wheel: Device has a directional wheel(s) for navigation (uncommon).
Platform Version (API level)	Examples: v3 v4 v7 etc.	The API level supported by the device. For example, v1 for API level 1 (devices with Android 1.0 or higher) and v4 for API level 4 (devices with Android 1.6 or higher **Caution:** Android 1.5 and 1.6 only match resources with this qualifier when it exactly matches the platform version.

Appendix B: Icon naming conventions

Use common naming conventions for icon assets

Asset Type	Prefix	Example
Icons	ic_	ic_star.png
Launcher icons	ic_launcher	ic_launcher_calendar.png
Menu icons and Action Bar icons	ic_menu	ic_menu_archive.png
Status bar icons	ic_stat_notify	ic_stat_notify_msg.png
Tab icons	ic_tab	ic_tab_recent.png
Dialog icons	ic_dialog	ic_dialog_info.png